Hydrogen Technologies

Offering a wide-range coverage, this book provides fundamentals as well as the applied science and technology involved in the whole hydrogen value chain, including production, storage, transportation, and utilization. It discusses some challenges and opportunities for hydrogen to address energy demand and climate change issues.

Features:

- Discusses various technology pathways for manufacturing/producing hydrogen both directly (i.e., water splitting) and indirectly (i.e., gas, conversion of coal, and biomass).
- Covers techniques and technologies for transporting gaseous, liquid, solid, and other forms of hydrogen, including mobile and stationary modes as well as small- and large-scale forms of transportation.
- Offers techniques and technologies for storing hydrogen with emphasis on materials and physical and chemical characteristics.
- Describes hydrogen utilization in energy/energy conversion, industrial chemical, industrial agricultural, and transportation sectors.

This book is aimed at engineers and scientists working in the disciplines of energy, chemical, environmental, petroleum, petrochemical, and mechanical engineering.

Olayinka I. Ogunsola served as Senior Program Manager at the Department of Energy (DOE), Office of Oil and Natural Gas, where he managed the department's onshore unconventional oil and gas research and development portfolio. Prior to joining DOE in 2000, Dr. Ogunsola was a National Research Council senior research associate at the U.S. Department of Energy, National Energy Technology Laboratory (NETL), Morgantown, West Virginia.

Olubunmi M. Ogunsola is Senior Research Safety Compliance Specialist at Howard University. She is an engineer with significant experience in the energy, minerals, and environmental industry and is owner/president of Tryby Energy Minerals & Environmental Corporation (TEMEC).

Hydrogen Technologies

Production, Transportation, Storage, and Utilization

Olayinka I. Ogunsola and
Olubunmi M. Ogunsola

CRC Press is an imprint of the
Taylor & Francis Group, an **informa** business

Designed cover image: © Shutterstock, Macrovector

First edition published 2025
by CRC Press
2385 NW Executive Center Drive, Suite 320, Boca Raton FL 33431

and by CRC Press
4 Park Square, Milton Park, Abingdon, Oxon, OX14 4RN

CRC Press is an imprint of Taylor & Francis Group, LLC

ISBN: 978-1-032-39071-0 (hbk)
ISBN: 978-1-032-39073-4 (pbk)
ISBN: 978-1-003-34828-3 (ebk)

DOI: 10.1201/9781003348283

Typeset in Times
by Newgen Publishing UK

Contents

Contents

Preface

The wheels of any nation's economy are primarily driven by energy. As such, the economic growth of a country is directly related to the availability, environmental sustainability, security, and affordability of its energy resources. While the demand for fossil energy continues to rise, production of domestic fossil fuels, particularly petroleum, is deemed to be incapable of meeting the needs of a growing economy. In addition, these fossil energy resources are plagued with their propensity to impact the environment (such as carbon dioxide emission, which can consequently result in climate change) both during their development and utilization. It is, therefore, essential to have a secure and diverse supply of affordable and environmentally friendly energy resources to achieve sustainable economic growth. Hydrogen, which does not contain carbon and can be produced from various sources (including renewables), seems to have potential to provide meaningful contribution toward addressing these issues.

While hydrogen is the most abundant element in the universe, it, however, does not occur on Earth in its molecular form and must hence be produced from other hydrogen-containing primary energy sources, such as fossil fuels (natural gas and coal), and renewable sources (biomass and water).

This book provides information on the fundamentals of technologies for producing, transporting, storing, and utilizing hydrogen. The book, which begins with an introductory chapter, covers fundamentals of hydrogen production from natural gas and coal via chemical and thermal technological pathways (such as steam reforming, autothermal reforming, partial oxidation, pyrolysis, and gasification), water electrolysis, and plasma electrolysis; transportation; storage; and various utilization methods, including industrial manufacturing sector (chemicals, iron and steel, power generation, synthetic fuel production, oil refining and upgrading, and transportation, such as in cars, trucks, rail, and aviation).

This book also discusses the pertinent properties of hydrogen that may influence the ways in which it is produced and used. The authors anticipate that this book will be particularly useful for hydrogen/hydrogen-based fuels production and application scientists, technologists, engineers, and plant designers and operators, as well as for researchers, professionals, technology developers, and policymakers. The book is also a valuable resource for continuing education, undergraduate, and graduate courses in hydrogen and hydrogen-based fuel programs.

1

Introduction

1.1 Overview

The word "hydrogen" comes from the Latin word "Hydrogenium". It can also be traced to the Greek words "Hydro" which means "water" and "gen" from "Genos" meaning "to create". Hydrogen is the most abundant element in the universe. The sun and other stars are composed largely of hydrogen. Hydrogen is the simplest non-metallic gaseous chemical element. Astronomers estimate that 90 percent of the atoms in the universe are hydrogen atoms. Hydrogen is a colorless, odorless, tasteless, flammable gas. It is the lightest of the known elements (Stwertka, 1996), with atomic number of 1. When combusted, it generates heat – that can be used for processes of heat and energy conversion – and water/steam (an environmentally safe product of combustion).

While hydrogen was first recognized by a British chemist, Henry Cavendish in 1766 (Emsley, 2001), and first named by Antoine Lavoisier in 1783 (Stwertka, 1996 and Van Nostrand's Encyclopedia of Chemistry, 2005), in 1671, Robert Boyle discovered and described the production of hydrogen gas from a reaction between iron filings and dilute acids (Winter, 2007). Hydrogen is known to be a discrete substance (Al-Khalili, 2010) – a substance that keeps its bonding interactions very quiet and straightforward, and with no floating electron clouds binding/bonding/banding with any, all and sundry atoms that just happen to float by.

While it is the most abundant element in the universe, molecular hydrogen does not occur on Earth and must, therefore, be produced from other hydrogen-containing primary energy sources, such as fossil fuels (natural gas and coal), biomass, water, and other renewables. Naturally occurring hydrogen is generated continuously from a variety of natural sources. There are many known hydrogen emergences on mid-ocean ridges. The main source of natural hydrogen is ultramafic rocks that have undergone serpentinization. Other natural processes of generating hydrogen have, however, been identified. According to a Wikipedia article (2022), these natural processes include:

- Degassing of deep hydrogen from the Earth's crust and mantle
- Reaction of water with ultrabasic rocks (serpentinization)

DOI: 10.1201/9781003348283-1

- Contact of water with reducing agents in the Earth's mantle
- Interaction of water with freshly exposed rock surfaces (weathering)
- Decomposition of hydroxyl ions in the structure of minerals
- Natural radiolysis of water
- Decomposition of organic matter
- Biological activity

A recent comprehensive review of natural hydrogen by Zgonnik (2020) concludes that molecular hydrogen is much more widespread in nature than was previously thought and that a deep-seated origin of hydrogen is probably the most likely explanation for its abundance in nature.

The notion of utilizing hydrogen as an energy carrier became significantly noticeable after the 1974 worldwide energy crisis (Bailieux, 1981; Noor and Siddiqi, 2010; Mazloomi and Gomes, 2012). Current increased interest in hydrogen production is a result of its (hydrogen) attributes compared with other fuels. Hydrogen is known to be a clean energy carrier mainly because of its carbon-free nature (Jovan and Dolac, 2020; Kulagin and Grushevenko, 2020; Bespalko and Mizeraczyk, 2022). Hydrogen has the highest specific energy density of about 121 MJ/kg (Rivard et al., 2019; Bespalko and Mizeraczyk, 2022). It also has the potential for a better and more reliable way of converting renewable electricity into renewable hydrogen for use in fuel cells, vehicles, and industrial manufacturing applications (Jovan and Dolac 2020; Rosen and Koohi-Fayegh, 2016; Bespalko and Mizeraczyk, 2022).

As mentioned earlier, man-made hydrogen (H_2) can be generated by extracting it from another compound, such as water (H_2O) or from fossil fuels (mainly natural gas and coal), nuclear energy, and from renewable (wind, solar, geothermal, plastics, biomass, and waste streams [municipal solid wastes and landfill gas]). While this book presents some insights on fundamentals and technologies of hydrogen development, details of the various methods of producing hydrogen and the several research efforts being conducted towards its (hydrogen) development can also be found elsewhere (Kalamaras and Eftathiou, 2013; Bespalko and Mizeraczyk, 2022; IEA, 2019; U.S. Department of Energy, 2019, 2020). Hydrogen then acts as an energy carrier and storage device, helping to generate electricity or heat when combined with oxygen or air through a fuel cell or when combusted directly. The main production pathways for hydrogen are through thermo-chemical processes, such as reforming (steam reforming, auto-thermal reforming, and partial oxidation), gasification, pyrolysis – heating of carbonaceous materials in the absence of oxidant – and by electrolytic process (electrolysis of water or water splitting). As outlined by Bespalko and Mizeraczyk (2022), the advantages of the various technological pathways for producing hydrogen include the following:

- **Natural gas steam reforming** – It is the most commercially used method, available as either a large-scale, centralized facility or a small-scale distributed unit (U.S. Department of Energy Hydrogen Plan, 2020), and with relatively high H_2/carbon dioxide ratio.

- **Partial oxidation** – It has fast reaction kinetics, requires a small reactor volume, and does not require external source of heat.
- **Electrolysis** – Does not emit CO_2 and it is a low-temperature operation that utilizes renewable energy source.
- **Photobiological methods** – The process uses microorganisms to convert solar energy into pure and clean hydrogen gas from water and sunlight, as the feedstock and energy source, respectively. The photobiological process is also amenable to producing other value-added products such as lactic acid, butyric acid, and acetic acid through photosynthetic reactions under anaerobic conditions. Under aerobic conditions, the process can be used to sequester CO_2 by converting it into biomass, which is subsequently converted to hydrogen under the anaerobic condition.
- **Thermochemical process** – The thermochemical process of producing hydrogen, which is powered by renewable heat and electricity, can be operated at significantly low temperature, thereby emitting no CO_2.

However, these processes are plagued with some drawbacks, which include the following, according to Bespalko and Mizeraczyk, (2022), and as discussed in the IEA (2019) and U.S. Department of Energy (2019, 2020) reports:

- **Stem methane reforming process** – The process requires a significant external energy source because of the endothermic nature of the underlying basic reactions involved. It is a catalytic process that requires large reactor size, a gas separation system, and it operates at high temperature, which makes it relatively expensive. It is a CO_2 emitting process because of its use of natural gas as a feedstock.
- **Partial oxidation** – Since the process utilizes hydrocarbon as the feedstock, it is, therefore prone to CO_2 emission. It also has a low H_2/CO_2 ratio and low H_2/CH_4 ratio.
- **Electrolysis** – Electricity is needed to power electrolyzer and may emit CO_2 if the electric power is produced by carbon-containing fuel. The process typically requires the use of catalysts to enhance hydrogen evolution reaction.
- **Thermochemical methods** – The system can be complex brought about by a high number of reactions required to break the water molecule. High operating temperature is required for the thermochemical reactions, and the use of catalysts is necessary in some instances.

Just like energy, hydrogen utilization has a long history. As such, hydrogen has been a critical part of the energy industry since the middle of the 1900s, when its use became popular in the oil refining industry (IEA, 2019). According to the 2019 International Energy Agency (IEA) report, the first internal combustion engines in the

1800s were fueled with hydrogen. It was also used to lift balloons and airships in the 18th and 19th centuries (IEA, 2019). Hydrogen also contributed to the exploration of the moon in the 1960s (IEA, 2019). Today, hydrogen use has found its way across numerous sectors, including energy conversion systems (such as power generation), fuel processing and upgrading (e.g., refining), industrial feedstock (such as in chemical, food, pharmaceuticals, metal manufacturing, fertilizer, etc.), commercial, transportation (engines), etc. As also indicated in the IEA (2019) report, global demand for pure hydrogen is estimated at 70 million metric tons per year, while the United States consumption is about 10 million metric tons annually, which is approximately 1 percent of U.S. energy consumption (DOE Hydrogen Program Plan, 2020; Global hydrogen demand is expected to continue to grow as new and future areas of hydrogen applications are expected to emerge and expand, as well as growth in existing areas. According to U.S. Department of Energy (2020) Hydrogen Economy Strategy report, it is predicted that the hydrogen markets could grow 10–20 times by 2050 from their current level. This prediction will depend on the extent of hydrogen use in the economy.

There are four main areas in which hydrogen is currently commonly used – transportation (cars, buses, and trucks), industrial manufacturing (chemical, oil refining, and upgrading), power generation, and integrated systems (U.S. Department of Energy Hydrogen Program Plan, 2020). In addition to the current areas of usage, the emerging future major areas of hydrogen utilization include rail, ocean-going vehicles, and aviation (in the transportation sector), iron and steel, cement, chemical, biofuels/synthetic fuels industries (in the industrial manufacturing sector), fuel cell power generation, hydrogen-fueled combustion system (boiler, furnace) for power and process heat generation, long-term energy storage, hydrogen–nuclear hybrid system, integrated gas–coal–hydrogen hybrid system, and carbon capture utilization and storage (CCUS), according to the IEA (2019) and the U.S. Department of Energy Hydrogen Program Plan (2020). As such, the need to understand the impacts of the fundamentals associated with its various production and utilization technology pathways cannot be over-emphasized. The possibility of a significant increase in expansion of hydrogen application in the future has attracted (and will continue to attract) interest and investments from government and industry, which will increase market potential for hydrogen production and utilization technologies both in the United States and worldwide (U.S. Department of Energy Hydrogen Program Plan, 2020; Hydrogen Council Report, 2017; and US Hydrogen Study.Org., 2019). In addition to its direct use in a wide range of applications, hydrogen can be used to store, move, and deliver low- or non-carbon-laden energy to its utilization destination (U.S. Department of Energy Hydrogen Plan, 2020). Hydrogen can be stored in all three phases or forms (liquid, gas, or solid).

The current hydrogen demand/consumption is being met through production from fossil fuels, mainly and internationally from natural gas, which accounts for about 95 percent and about 76 percent of the U.S. and world hydrogen production, respectively. This is followed by coal-derived hydrogen, and lastly by electrolytic hydrogen, which accounts for about 2 and 1 percent of the global and U.S. hydrogen production, respectively (IEA, 2019; DOE Hydrogen Program Plan 2020). As a result of the fact that hydrogen production today is largely from fossil fuels, the role of hydrogen in addressing climate change may be questionable unless these technologies are equipped or integrated with cost-effective CCUS technologies.

While hydrogen can be used in a wide range of application areas (as enumerated earlier), and it is currently the only zero-emission technology capable of producing the extreme heat required by the various end uses (such as to produce cement, steel, glass, and other industrial materials), carbon emission, however, can occur from other sections of the value chain. These stages include production, storage, transportation, and some aspects of utilization. Carbon emission can occur during hydrogen production stage, especially if it is produced from fossil fuels (mainly coal and natural gas) and other carbon-containing feedstock, such as biomass, municipal solid waste, and landfill gas. It was reported by IEA (2019) that hydrogen production is responsible for emission of about 830 million metric tons of carbon dioxide per year ($MMtCO_2$/year). Carbon emission can be reduced or eliminated during hydrogen production if the process is integrated or combined with carbon capture and storage (CCS) technology system or if the feedstock is de-carbonized, as mentioned. These technologies are being developed at the various industry and government entities (including the U.S. Department of Energy).

1.2 Occurrence, Nature, and Types of Hydrogen

1.2.1 Occurrence and Origin

Hydrogen can be found in the free state of some volcanic gases as well as the outer atmospheres of the sun and other stars in the universe, accounting for about half of the mass of the sun and several other stars. Jupiter and Saturn are predominantly hydrogen-based planets. The nuclear fusion of hydrogen atoms, which releases a huge quantity of energy, is a consequence of the extremely high temperatures of the sun and stars. The emergence of neutral hydrogen atoms throughout the universe occurred about 370,000 years later during the recombination epoch, when the plasma had cooled enough for electrons to remain bound to protons (Tanabashi, 2018).

Hydrogen has been reported to exist in the following three different forms (Zgonnik, 2020):

- Hydrogen that exists as a free gas. It is, however, not found in our atmosphere because the earth's gravitational pull is not strong enough to hold light hydrogen molecules.
- Hydrogen as inclusions in other materials, such as Earth' crust, rocks, minerals, including coal, petroleum, oil, and natural gas. It makes up 15.4 percent of the earth's crust and seas. It is tenth in the order of abundance in crystal rocks.

- Hydrogen as dissolved gas or in combination with other materials, such as oxygen in the form of water; and in combination with carbon, nitrogen, and halogens in the form of organic matter in plant and animal tissues, carbohydrates, proteins, and other substances.

Most of the hydrogen on Earth exists in molecular forms such as water and organic compounds (i.e., second category of form). More details on the occurrence of natural hydrogen can be found in a review paper published by Zgonnik (2020). The abundance of occluded gases in rocks and minerals is covered elsewhere (Petersilie, 1964; Ikorsky et al., 1992; Nivin, 2019). Protium, deuterium, and tritium are the three naturally occurring hydrogen isotopes. More discussion on hydrogen isotopes is given in Chapter 2.

1.2.2 Types of Hydrogen

Hydrogen, in its pure state, is known to be odorless, invisible to naked eyes, and colorless. But it is now described with a range of colors, which are based on its (hydrogen) source and method of production (Sen et al., 2022). There are seven different colors (black, gray, green, blue, turquoise, pink, and white) associated with hydrogen based on the type of feedstock and production technology used. While a detailed description of the various colors assigned to hydrogen is provided by Sen et al. (2022), they are defined as follows:

- **Black/brown hydrogen** – This is defined as hydrogen produced through coal gasification, which leads to generation of CO_2. Brown hydrogen market is expected to rise with time in the future (Allied Market Research, 2021).
- **Gray hydrogen** – This is hydrogen produced through natural gas reforming and is responsible for about 6 percent of global natural gas consumption (Farmer, 2020). Gray hydrogen can also be produced through the partial oxidation of refinery residues. Gray hydrogen is currently the cheapest source of manufacturing industrial hydrogen. It is a CO_2-emitting process.
- **Green hydrogen** – This is referred to as hydrogen produced via electrolysis of water, powered with renewable energy sources (such as solar or wind), and with no greenhouse gas emission. Green hydrogen is expensive, costing about $ 6/kg, according to the National Hydrogen Council.
- **Blue hydrogen** – This is when it is primarily produced from natural gas, using steam methane reforming (SMR) or autothermal reforming (ATR) technology integrated with CCUS technology.
- **Turquoise hydrogen** – Is defined as hydrogen produced via natural gas pyrolysis process. It is a low carbon intensity process, even more so if the heat energy used for the pyrolysis is provided by a renewable source.
- **Pink hydrogen** – Is defined as hydrogen manufactured basically through a nuclear-energy-powered electrolysis process. The primary goal of pink

hydrogen production technology is to lower the cost of clean hydrogen (U.S. Department of Energy, 2019). It is, however, plagued with safety issues associated with radioactive waste production.

- **White hydrogen** – This is defined as naturally occurring hydrogen, (such as that found in rock and geological formation, oceanic or continental crust, volcanic gas, geysers, hydrothermal systems). It is probably the cheapest solution to produce carbon-neutral hydrogen. According to Solar Impulse Foundation (2021), current estimates of the flux provide 23 million tons of hydrogen per year from all geologic sources combined.

However, it is worth noting that from all practical purposes, hydrogen is colorless regardless of its method and source of production.

1.3 Rationale and Driving Forces for Hydrogen

There are two main factors driving the reincarnation of hydrogen economy. These are (1) energy security and (2) climate change and environmental issues. The wheels of any nation's economy are primarily driven by energy. A country's economic growth is known to be directly related to its energy consumption. While the demand for fossil energy continues to rise, production of domestic fossil fuels, particularly petroleum, is deemed to be incapable of meeting the needs of a growing economy. It is, therefore, essential to have a secure supply of affordable energy to achieve sustainable economic growth. This is more so for the United States and other developed countries. According to the U.S. DOE Energy Information Administration (EIA) (2022) Annual Energy Outlook (AEO), petroleum and natural gas remain the most-consumed sources of energy in the United States through 2050.

Per the EIA (2022) AEO, the world's consumption of petroleum and other liquids is about 92 million barrels per day, which is predicted to rise to over 125 million barrels per day by 2050. United States' import of crude oil was 3.24 million barrels per day in 2021. This is predicted to increase to 3.97 million barrels per day by 2050 (EIA, 2022,Annual Energy Outlook, 2022). This is an increase of about 22 percent over a period of 38 years.

The other major driving force for use of hydrogen as fuel is the environmental challenge posed by fossil-fuel utilization, which results into increased emission of combustion-generated air pollutants, such as carbon dioxide (a greenhouse gas), carbon monoxide, nitrogen oxides (NO_x), volatile organic compounds (VOCs), and sulfur dioxide (SO_2). These pollutants can have a significant negative local, urban, regional, and global impact on human health and the environment, if not abated. Hydrogen, therefore, is a promising candidate for reducing these impacts, thereby providing a cleaner environment. Major sources of these emissions include transportation, industrial, commercial, and residential. Among these sectors, transportation and power generation sectors account for the largest fraction of the man-made air pollution in the United States.

Another role for hydrogen in reducing environmental impact is in climate change. The U.S. transportation and electric power sectors combined are responsible for three quarters of man-made emissions of carbon dioxide (CO_2). Global annual energy-related CO_2 emission is about 33.61 billion metric tons, as reported in the 2022 AEO (EIA, 2022). The world annual CO_2 emission is predicted to increase to about 43 billion metric tons per year by 2050 (EIA, 2022). The increase in projected energy consumption and energy-related CO_2 emission is a result of population and economic growth. Research and development efforts are being made at the government and various private research laboratories to better understand the causes of global climate change and develop cost-effective and efficient technologies to reduce the growth of greenhouse gas emissions. Hydrogen as an energy carrier has potential in providing a solution to eliminate or reduce man-made greenhouse gas emissions when carbon capture and storage technologies are integrated with the hydrogen production technologies especially when produced from low-carbon energy source. However, producing hydrogen from low-carbon energy sources is currently cost-prohibitive. According to IEA (2019), the cost of renewable electricity-derived hydrogen could decrease by 30 percent by 2030, which is expected to result from declining cost of renewables and hydrogen production scale up.

This book presents fundamental science and technology of hydrogen production, transportation and delivery, storage, and utilization. This chapter gives a general introduction on hydrogen, while the properties of hydrogen as they influence its methods of production and utilization are presented in Chapter 2. Chapters 3 through 5 describe the fundamentals and technologies for producing hydrogen via electrolysis, fossil fuels (coal and natural gas), and biomass, respectively. An account of the fundamentals and technologies for transporting and delivering hydrogen to various users is given in Chapter 6, while Chapter 7 covers technologies for its storage. The fundamentals and technologies of the primary pathways of utilizing hydrogen in the transportation, oil refining and upgrading, synthetic fuel production, and industrial manufacturing sectors are presented in Chapter 8.

References

Al-Khalili, J. (2010). Discovering the Element. Chemistry: A Volatile History. 25:40, A British Broadcasting Corporation Four (BBC 4) Presentation.

Allied Market Research. (2021). Brown Hydrogen Market to Reach $48.9 Bn, Globally, by 2030 at 4.7% CAGR: Presence of Abundant Reserves and Cheap Raw Materials Drives the Growth of the Global Brown Hydrogen Market.

Bailleux, C. (1981). Advanced Water Alkaline Water Electrolysis: A Two-year Running of a Test Plant. *Int. J. Hydrogen Energy*, 6, 46–71.

Bespalko, S. and Mizeraczyk, J. (2022). Overview of Hydrogenation by Plasma-Driven Solution Electrolysis. *Energies*, 15(20), 7508.

Emsley, J. (2001). *Nature's Building Blocks*. Oxford University Press: Oxford, England, pp. 183–191. ISBN 978-0-19-850341-5.

Farmer, M. (2020). What Color Is Your Hydrogen? A Power Technology Jargon-Buster: All Hydrogen Burns the Same, But the Different Methods of Producing it Have Produced Colorful Nicknames. Power Technology.

Hydrogen Council Report. (2017). Hydrogen Scaling Up. A Sustainable Pathway for the Global Energy Transition. https://hydrogencouncil.com/wp-content/uploads/2017/11/Hydrogen-scaling-up-Hydrogen-Council.pdf.

Ikorsky, S.V., Nivin, V.A., and Pripachkin, V.A. (1992). *Gas Geochemistry of Endogenic Formations*. Nauka: St. Petersburg, Russia.

International Energy Agency (IEA). (2019). The Future of Hydrogen – Seizing Today's Opportunities. A Report Prepared by IEA for the G20.

Jovan, D.J. and Dolanc, G. (2020). Can Green Hydrogen Production Be Economically Viable under Current Market Conditions. *Energies*, 13, 6599.

Kulagin, V.A. and Grushevenko, D.A. (2020). Will Hydrogen Be Able to Become the Fuel of the Future? *Therm. Eng.*, 67, 189–201.

Mazloomi, K. and Gomes, C. (2012). Hydrogen as an Energy Carrier: Prospects and Challenges. *Renew. Sustain. Energy Rev.*, 16, 3024–3033.

Nivin, V.A. (2019). Occurrence Forms, Composition, Distribution, Origin and Potential Hazard of Natural Hydrogen–Hydrocarbon Gases in Ore Deposits of the Khibiny and Lovozero Massifs: A Review. *Minerals*, 9(9), 535; https://doi.org/10.3390/min9090535.

Noor, S. and Siddiqi, M.W. (2010). Energy Consumption and Economic Growth in South Asian Countries: A Co-Integrated Panel Analysis. *Proc. World Acad. Sci. Eng. Technol.*, 67, 251–256.

Petersilie, I.A. (1964). *Geology and Geochemistry of Natural Gases and Disperse Bitumens of Some Geological Formations of the Kola Peninsula*. Nauka: Moscow, Russia.

Rivard, E., Trudeau, M., and Zaghib, K. (2019). Hydrogen Storage for Mobility: A Review. *Materials*, 12, 1973.

Rosen, M.A. and Koohi-Fayegh, S. (2016). The Prospects for Hydrogen as an Energy Carrier: An Overview of Hydrogen Energy and Hydrogen Energy Systems. *Energy Ecol. Environ.*, 1, 10–29.

Sen, S., Mani Bansal, M., Syed Razavi, S., and Khan, A.S. (2022). The Color Palette of the Colorless Hydrogen. The Way Ahead. https://jpt.spe.org/twa/the-color-palette-of-the-colorless-hydrogen.

Solar Impulse Foundation. (2021). White Hydrogen: Drilling for Natural Hydrogen. https://solarimpulse.com/solutions-explorer/white-hydrogen.

Stwertka, A. (1996). *A Guide to the Elements*. Oxford University Press: Oxford, England, pp. 16–21. ISBN 978-0-19-508083-4.

Tanabashi, M. (2018). Big-Bang Cosmology. Archived 29 June 2021 at the Wayback Machine (Revised September 2017) by K.A. Olive and J.A. Peacock. Chapter 21.4.1, p. 358.

U.S. Department of Energy. (2020). Hydrogen Economy Strategy: Fossil Energy Enabling a Hydrogen Economy.

U.S. Department of Energy Hydrogen Program Plan. (2020). DOE/EE – 2128.

U.S. Department of Energy, Energy Information Administration. (2022). Annual Energy Outlook 2022.

U.S. Department of Energy Information Administration. (March 19, 2019). Today in Energy. www.eia.gov/todayinenergy/detail.php?id=38773

US Hydrogen Study.Org. (December 2019). E4 tech. The Fuel Cell Industry Review, 2019.

Van Nostrand's Encyclopedia of Chemistry. (2005). *Hydrogen.* Wiley-Interscience: Hoboken, NJ, pp. 797–799. ISBN 978-0-471-61525-5

Winter, M. (2007). Hydrogen: Historical Information. WebElements Ltd. Archived from the original on 10 April 2008. Retrieved 5 February 2008. https://en.Wikipedia.org

Zgonnik, V. (2020). The Occurrence and Geoscience of Natural Hydrogen: A Comprehensive Review. *Elsevier Earth-Science Rev.*, 203, 103–140.

2

Properties of Hydrogen

2.1 Introduction

Although hydrogen consists of minute percentage of Earth's crust by weight, it is, however, the most abundant substance in the universe (Dunn, 2002; Momirlan and Veziroglu, 2005), occurring in vast quantities as part of the water in oceans, ice packs, rivers, lakes, and the atmosphere. It is the lightest and has the simplest atomic structure of the chemical elements (Stwertka, 1996; Grolier, Inc., 1991). It is a colorless, tasteless, odorless element (Momirlan and Veziroglu, 2005; Abdel-Aal et al., 2005), and is a non-toxic, flammable substance with a symbol **H** and atomic number 1 (Stwertka, 1996). The name hydrogen has its origin from the Greek words – *hydro* (meaning water) and *genes* (forming, maker of water) – which is its characteristic chemical property of forming water (H_2O) when it is reacted with oxygen. Under standard conditions, hydrogen is a gas of diatomic molecules having the formula H_2. The density of gaseous hydrogen is very low (Verhelst and Wallmer, 2009; Mazloomi and Gomes, 2012). Hydrogen is characteristically bonded with other materials or elements such as carbon and oxygen and does not typically exist on its own in consumable scale (Balat, 2008; Mazloomi and Gomes, 2012). Hydrogen is also known to be present in all animal and vegetable tissue petroleum, contained in almost all carbon compounds, and forms many compounds with most elements.

As alluded to in Chapter 1, hydrogen gas was first artificially produced in the early 16th century through the reaction of acids on metals and was first recognized to be a discrete substance by Henry Cavendish around mid-1700 (Al-Khalili, 2010). This chapter describes the various properties of hydrogen (atomic structure, physical, chemical, combustion, etc.) with some reference to their roles in its (hydrogen) use where appropriate.

2.2 Atomic Properties

A hydrogen atom is a chemical element of hydrogen, and it is electrically neutral containing a single proton and a single electron that are positively and negatively charged, respectively, both of which are bound to the nucleus by a Coulomb force. According to Mazloomi and Gomes (2012), atomic hydrogen accounts for

DOI: 10.1201/9781003348283-2

approximately three quarters of the baryonic mass of the universe. Hydrogen, which was formally positioned in front of Group iA in the periodic table (the alkali metals), is now considered to be a unique element (Grolier, Inc., 1991). The presence of atomic hydrogen on its own on Earth is very rare, but rather it tends to combine with other atoms contained in other compounds, or with another hydrogen atom to form diatomic, molecular hydrogen. A brief description of some important atomic properties of hydrogen is given later. For this book, the properties of interest are classified into three main groups – atomic properties, chemical and physical properties, and reactivity/combustion properties. The atomic properties discussed here include oxidation state, ionization potential, covalent radius, Van der Waals radius, and isotope.

2.2.1 Oxidation State

As revealed by atomic spectroscopy, a discrete infinite set of states exists in hydrogen. The oxidation state of hydrogen is the hypothetical charge of hydrogen atom if all its bonds to other atoms are fully ionic (or the degree of loss of electrons). Hydrogen oxidation state is ±1 (i.e., an amphoteric oxide), as reported by Jolly (2024).

2.2.2 Ionization Energy

Ionization energy is the minimum energy required to remove the most loosely bound electron of an isolated gaseous atom, positive ion, or molecule, which can be quantified by Equation [2.1],

$$X\,(g) + \text{energy} \longrightarrow X^{+}\,(g) + e- \qquad [2.1]$$

where X is any atom or molecule, X^{+} is the resultant ion when the original atom was stripped of a single electron, and e– is the removed electron (Miessler and Tarr, 1999). It will require about 1312 kJ to ionize one mole of hydrogen. This is lower than the 1735 kJ of energy needed to ionize one mole of methane (Jolly, 2024). Therefore, less amount of energy is required to ionize hydrogen compared with methane.

2.2.3 Covalent Radius

Covalent radius, which is typically measured either in picometers (10^{-12} m) or angstroms, Å (10^{-10} m), is a measure of the size of an atom that forms part of one covalent bond. From basic chemistry, the sum of the two covalent radii should be equal to the covalent bond length between two atoms. Methods used in measuring the bond lengths include X-ray diffraction, neutron diffraction on molecular crystals, and rotational spectroscopy with varying degrees of accuracy. As published in the Cambridge Structural Database (Cordero, 2008), covalent radius of hydrogen atom has a 31±5 picometers, which is about half the covalent radius of an oxygen atom (about 66 picometers).

2.2.4 Van der Waals Radius

The van der Waals radius of an atom is the radius of an imaginary hard sphere representing the distance of closest approach for another atom. It is named after Johannes Diderik van der Waals, winner of the 1910 Nobel Prize in Physics, as he was the first to recognize that atoms were not simply points and to demonstrate the physical consequences of their size through van der Waals equation of state. It may be determined from the mechanical properties of gases, the critical point, measurements of atomic spacing between pairs of unbonded atoms in crystals, or from measurements of electrical or optical properties.

2.2.5 Hydrogen Isotopes

There are three naturally occurring isotopes of hydrogen – ^{1}H (with mass number of 1), ^{2}H (with mass number of 2), and ^{3}H (with mass number of 3), according to Quaching (2005) and Stojic et al. (1994). There are also other isotopes of hydrogen – ^{4}H to ^{7}H – with highly unstable nuclei that have been synthesized in the laboratory but not observed in nature (Gurov et al., 2004; Korsheninnikov et al., 2003).

Hydrogen isotope of mass number of 1 is the most abundant and is generally referred to as protium. The ^{2}H isotope, which has a nucleus of one proton and one neutron, is known as deuterium or heavy hydrogen (D), accounts for 1.56 parts per million (ppm) of the ordinary mixture of hydrogen, while tritium (T or ^{3}H), with one proton and two neutrons in each nucleus, accounts for about 10^{-15} to 10^{-16} percent of hydrogen. Each hydrogen isotope has distinct properties.

As mentioned, ^{1}H, which is also referred to as protium is the most common and most prolific hydrogen isotope, accounting for more than 99.98 percent of hydrogen (Urey et al., 1933). The absence of neutrons in ^{1}H isotope makes it unique among all stable isotopes.

^{2}H (deuterium) is another stable hydrogen isotope, containing one proton and one neutron in the nucleus. Deuterium is neither toxic nor radioactive, and deuterium-rich water is referred to as heavy water, which is commonly used for cooling nuclear reactors. It is also used in solvents for ^{1}H-NMR (nuclear magnetic resonance) spectroscopy (Broad, 1991), and has also been shown to have potential use as fuel for commercial nuclear fusion (Quaching, 2005).

^{3}H (tritium), which is radioactive, decaying into helium-3 through beta decay with a half-life of 12.32 years, contains one proton and two neutrons in its nucleus, according to Miessler and Tarr (1999). Its radioactive characteristic enables it to be used in luminous paint, thereby enhancing the visibility of the paint when used for data displays, such as in hands and dial-markers of watches. Other areas of using tritium include fusion reactions (Nave, 2006), as a tracer in isotope geochemistry (Kendall and Caldwell, 1998) and in specialized self-powered lightening devices, as well as in chemical and biological labeling experiments as a radiolabel (Holte et al., 2004).

2.3 Physical and Chemical Properties

Hydrogen belongs to the hexagonal crystal family which consists of the 12-point groups with at least one of their space groups having the hexagonal lattice as underlying lattice and is the union of the hexagonal crystal system and the trigonal crystal system (Dana and Hurlbut, 1959).

Hydrogen is gas at standard temperature and pressure. Because of the weak forces of attraction between hydrogen molecules, the melting and boiling points of hydrogen are extremely low (–259 and –253°C, respectively), as revealed by basic chemistry. The presence of these weak inter-molecular forces is a consequence of the tendency of hydrogen temperature to rise when it expands from high to low pressure at room temperature contrary to that of most other gases. Thermodynamically, it means that repulsive forces are greater than the attractive forces between hydrogen molecules at room temperature, which may otherwise cool the hydrogen during expansion. This cooling effect is much more pronounced at temperatures below that of −196°C, the effect of which could be utilized to achieve hydrogen liquefaction temperature. Briefly described later in this section are some other physical and chemical properties (electronegativity, bond dissociation energy, density, melting point, and boiling point) of hydrogen. Reactivity and combustion properties of hydrogen are presented in the next section.

2.3.1 Electronegativity

As defined by the International Union of Pure and Applied Chemistry (IUPAC) Compendiums of Chemical Terminology (1997), electronegativity is the propensity of an atom (such as that of hydrogen) in attracting shared electrons (or electron density) when forming a chemical bond. It serves as a simple way to quantitatively estimate the bond energy and is used to describe how strongly different atoms attract electrons. Hydrogen has high electronegativity, with a value of 2.2 (as learned from basic chemistry), and was chosen as the reference, as it forms covalent bonds with a large variety of elements. The more bonded electronegative an atom is, the more electrons it pulls towards itself, and the more reactive the atom is. The electronegativity values are non-dimensional (i.e., they have no units) and are reported relative to the standard reference, hydrogen. The high electronegativity value possessed by hydrogen is because it is a very small atom with a strong pull from its nuclear charge on the electrons in a bond. The polarity of a molecule because of the electronegativity of the atoms that make up the molecule is, therefore, important in the reactivity of the molecule.

2.3.2 Dissociation Energy

The bond dissociation energy is the energy required to break a bond and form two atomic or molecular fragments, each with one electron of the original shared pair. Hence, a very stable bond has a large bond dissociation energy – more energy must be provided to break the bond. From basic chemistry, about 104 kcal is required to dissociate one mole of hydrogen. This is the minimum amount of energy required to break the bond that holds together the atoms in the molecule. At that point, one molecule of hydrogen dissociates into two atoms ($H_2 \rightarrow 2H$) when that amount of energy is applied, as learned from basic chemistry.

Generally, the higher the bond energy, the shorter the length of the bond. The high dissociation energy manifested by hydrogen is because of its relatively small atom and short bond length, thereby having a very strong attraction between the atoms, consequently, a high bond dissociation enthalpy.

2.3.3 Density

Density is defined as mass per unit volume of a substance. It varies with temperature and pressure, and hence it varies with the state or phase of the substance. Although hydrogen has the highest energy per mass of any fuel, its low density at ambient conditions results in a low energy per unit volume. Density of gaseous hydrogen is about 0.09 g/cm^3 at standard temperature and pressure, while density of liquid hydrogen (about 0.07 g/cm^3) is lower than that of gaseous hydrogen. As such, generally, the density of hydrogen is relatively lower than that of other gaseous fuels. Consequently, the energy content of hydrogen is a lot lower than that of most other fuels and energy carriers. Therefore, storing or using hydrogen at atmospheric pressure and temperature requires a substantial amount of space. Also, since the density of liquid hydrogen is lower than that of gaseous hydrogen, the storage density of hydrogen is significantly increased when it is converted to a liquid state, making it possible for a much larger volume of liquid hydrogen to be transported per trailer than compressed gaseous hydrogen, as mentioned in Chapter 6.

By virtue of hydrogen's low density, it is very easy for other compounds to easily diffuse through hydrogen molecules, thereby making hydrogen react easily and rendering it an excellent heat conductor. However, the low density of hydrogen plagues it with two problems – very large storage volume of hydrogen is required for an adequate driving range and reduced engine power output because of the energy density of hydrogen–air mixture.

2.3.4 Melting Point and Boiling Point

The melting point (or liquefaction point) of a substance is the temperature at which the substance changes its state from solid to liquid. At this point, the solid and liquid phase co-exist in equilibrium. The melting point of a substance depends on pressure. Hydrogen has a relatively low melting point (about –259°C). This is attributed to the fact that hydrogen is made up of small, light molecules bonded together by relatively strong intermolecular forces. As a result, the melting point and boiling point of a substance or compound can be increased when it contains hydrogen. This is because hydrogen bonds have strong intermolecular attractive forces that create stable molecules. Hydrogen bond resilience time functions and autocorrelation functions show faster decay at higher temperatures. The breakage of hydrogen bonds appears to be primarily temperature dependent, although the frequency of bond breakage is slightly higher at the higher density. Hydrogen gas is difficult to condense to a liquid because of its extremely low molecular mass.

The boiling point of a substance is the temperature at which the vapor pressure of a liquid equals the pressure surrounding the liquid (Goldberg, 1988) and the liquid changes into vapor (Theodore et al., 1999), at which point the substance boils. The lower the boiling point of a substance, the more easily it boils. The practical application of this is manifested during storage of the substance. In the case of hydrogen, which can be transported in tanks either in gaseous or liquid form, some of the hydrogen molecules can boil off, resulting in its loss during transportation and/or storage. Hence, tanks for transporting and storing hydrogen, which have a relatively low boiling point (–253°C), must be designed with this property in mind.

2.4 Hydrogen Reactivity and Combustion Properties

2.4.1 Reactivity

Hydrogen is unreactive compared with diatomic elements such as halogens or oxygen. Its low reactive characteristics can be explained thermodynamically by its very strong H–H bond (with a bond dissociation energy of 435.7 kJ/mol) (Lide, 2006) and kinetically its no or weak polar nature. It spontaneously reacts with some halogens, notably chlorine and fluorine, to form halides (hydrogen chloride and hydrogen fluoride, respectively) (Clayton, 2003). The reaction for forming halides from hydrogen is represented by Equation [2.2]

$$X_2\,(g) + H_2\,(g)^+ \rightarrow 2HX\,(g) \qquad [2.2]$$

where X could be any halogen (such as chlorine, bromine, or fluorine).

Hydrogen also readily reacts with molecular nitrogen (N_2) to form ammonia (NH_3) at a temperature of about 400°C and a pressure of 200 atm in the presence of iron (Fe) as catalyst (Klerke et al., 2008). Hydrogen is non-metallic except at extremely high pressures at which point it readily forms a single covalent bond with most non-metallic elements, thereby forming compounds such as water and nearly all organic compounds as well as metals. Hence, hydrogen readily reacts with metals at a high temperature to produce hydrides. Hydrogen has also been found to reduce some metal oxides and metal ions in aqueous solutions into the appropriate metals. The equations representing the fundamental underlying reaction of hydrogen with nitrogen to form ammonia, metals to form hydrides, and the reduction of metal oxides and ions to form metals are shown in equations [2.3], [2.4], and [2.5], respectively.

$$3H_2\,(g) + N_2\,(g) \rightarrow 2NH_3\,(g) \quad [2.3]$$

$$Pd_2^{+}\,(aq) + H_2\,(g) \rightarrow 2H^{+}\,(aq) + Pd\,(s) \quad [2.4]$$

$$H_2\,(g) + 2M\,(g) \rightarrow 2MH\,(s) \quad [2.5]$$

where M is an alkali metal.

Hydrogen also finds its role in acid–base reactions which usually involve the exchange of protons between soluble molecules. Hydrogen can take the form of anion (negatively charged ions) where it is known as a hydride or as a cation (positively charged ions, H^+). The H^+ cation is a proton that behaves in aqueous solutions and in ionic compounds. These characteristic properties and behavior of hydrogen explain its role in reacting with metals to form hydride and in its ability to reduce metal oxides to metals, as described previously and as depicted by equations [2.3–2.5].

A disadvantage in the reactivity of hydrogen with metals is, however, manifested by the potential of hydrogen embrittlement occurring in steel pipes when used to transport hydrogen or hydrogen-containing gaseous fuels (Rogers, 1999). Many mechanisms have been proposed (Robertson et al., 2015) and evaluated as to the fundamental cause of hydrogen embrittlement (Bhadhesia, 2016), the results of which has led to the wide acceptance in recent years that hydrogen embrittlement is a complex process and that no single mechanism applies exclusively (Haiyang, 2009). Generally, the mechanisms for hydrogen embrittlement in steel pipelines include the formation of brittle hydrides, the creation of voids that can lead to high-pressure bubbles, enhanced decohesion at internal surfaces, and localized plasticity at crack tips that assist in the propagation of cracks (Robertson et al., 2015).

2.4.2 Combustion Properties

As mentioned in Section 2.1 and as will be discussed later in Chapter 8, hydrogen gas or molecular hydrogen is highly flammable when reacts with an oxidant such as air or oxygen. The governing equation of such reaction is represented by Equation [2.6]:

$$2H_2\,(g) + O_2\,(g) \rightarrow 2H_2O\,(l)\ 286\ kJ/mol \quad [2.6]$$

H_2/O_2 or H_2/air combustion readily occurs when conducted in appropriate mixture with air or oxygen and in the presence of metal catalysts.

Combustion behavior of hydrogen is determined primarily by various hydrogen properties which include a wide range of flammability, ignition energy, small quenching distance, high autoignition temperature, high flame speed at stoichiometric ratios, very low density, and high diffusivity. A more detailed list of pertinent combustion characteristics of hydrogen and those of compressed natural gas (CNG) and gasoline (for comparison) can be found elsewhere (Chatterjee et al., 2014; Mazloomi and Gomes, 2012).

Hydrogen combustion can also be achieved when mixed with oxygen. However, hydrogen–oxygen flames tend to emit ultraviolet light and are nearly invisible to the naked eye. This renders hydrogen use in combustion-related devices unsafe thereby requiring a leak and flame detector. However, hydrogen flames in other conditions are blue, resembling blue natural gas flames (Schefer et al., 2009). While the combustion of hydrogen to generate heat, power, electricity, and as transportation fuel does not emit carbon dioxide, combustion of hydrogen can, however, lead to the formation of thermal nitrogen oxides (NO_x), a criteria pollutant (Lewis, 2021), as a result of high flame temperature of hydrogen flame.

2.4.3 Flammability Limits

Flammability limits are the entire range of concentrations of a mixture of flammable vapor or gas (hydrogen in this case) in air over which a flame will occur and travel if the mixture is ignited. There are generally two limits – lower limit and upper limit. The lower flammability limit (LFL) is the lowest concentration of the flammable gas (i.e., hydrogen) in the mixture that can sustain a flame, while the upper flammable limit (UFL) is the highest concentration of the flammable component in the mixture, above which the mixture cannot burn, and the flame will not be sustained. It is worth noting that the wider flammability range (4–75 percent; Mazloomi and Gomes, 2012) exhibited by hydrogen enables hydrogen flame to be stable over a wide range of hydrogen/air ratio and under highly dilute conditions, thereby allowing for better operational flexibility, lower emission propensity, and a wide range of applications as a fuel (Chatterjee et al., 2014; Momirlan and Veziroglu, 2005).

Basically, the flammability limits of a fuel in an oxidant (air or oxygen) are the concentrations of the fuel that can produce and sustain a flame when mixed with an oxidant (air or oxygen). Another advantage of wide range of flammability limits exhibited by hydrogen, which allows engines or combustion devices fired by hydrogen to run on a lean mixture, is one in which the amount of fuel required is less than the theoretical stoichiometric ratio, or chemically ideal amount needed for combustion with a given amount of air, thereby resulting in low fuel consumption (Momirlan and Veziroglu, 2005).

2.4.4 Autoignition Temperature

The autoignition temperature of a fuel is the lowest temperature at which the fuel spontaneously ignites under normal atmospheric conditions without an external source of ignition, such as a flame or spark (Laurendeau and Glassman, 1971). It is also referred to as self-ignition temperature, spontaneous ignition temperature, or minimum ignition temperature. Autoignition temperature is the temperature required to provide the activation energy needed to sustain combustion. The autoignition temperature of hydrogen is about 858 K (Patnaik, 2007), which is higher than that of CNG and a lot higher than that of gasoline (Chatterjee et al., 2014). However, hydrogen flame luminosity is very low compared with that of natural gas and gasoline (Kothari et al., 2004). The combination of low flame luminosity and high ignition temperature characterized by hydrogen, therefore, makes it a safer fuel than natural gas and gasoline, as reported by Kothari et al. (2004).

2.4.5 Flame Velocity

Another important combustion property of hydrogen is the flame velocity or burning velocity, which is defined as the velocity at which unburned gases move through the combustion zone in the direction normal to the flame front. This property depicts the ability of the flame to remain stable during combustion. Hydrogen flame speed seems to be about four times higher than those of CNG and gasoline (Chatterjee et al., 2014). The high flame velocity exhibited by hydrogen coupled with its wide flammability limits enables hydrogen to be a very suitable fuel for internal combustion engines, jet engines, and gas turbines (Kothari et al., 2004).

2.4.6 Adiabatic Flame Temperature

Adiabatic flame temperature, which is referred to as the temperature of the products of combustion of a fuel after all chemical reactions have reached equilibrium and when no heat is allowed to escape (or enter) the combustor, is another useful combustion property of fuels such as hydrogen. It is useful in determining the energy released by the flame to the combustion products during combustion, which consequently leads to temperature rise. Each fuel has a unique adiabatic flame temperature for a given amount of oxidant. Adiabatic flame temperature is useful in the design of a combustion system and in the choice of the material of construction of the device. Adiabatic flame temperature of hydrogen is lower than that of gasoline, but higher than that of compressed natural gas (Kothari et al., 2004).

2.4.7 Fuel/Air Ratio

Air/fuel ratio or fuel/air ratio is the appropriate ratio of fuel and air that can initiate and sustain a flame or combustion. The minimum fuel/air ratio at which a complete combustion of a fuel is achieved is referred to as its stoichiometric ratio. This is about 34.5/1 for hydrogen–air combustion (Kothari et al., 2004). An incomplete combustion, which can cause emission of combustion-generated air pollutants, results when the combustion system is operated at a fuel/air ratio higher than the stoichiometric ratio, while operating at a fuel/air ratio less than the stoichiometric fuel/air ratio (referred to as fuel lean condition) can lead to the presence of excess oxygen or excess air in the products of combustion and consequently impact the energy release. Hydrogen is also known to have high diffusivity, which is about four times that of gasoline, thereby enhancing its mixability with air during combustion (Chatterjee et al., 2014), and consequently improving its combustion performance in engines.

2.4.8 Heating Value

Also of importance to hydrogen combustion is the low heating value (LHV), which is referred to as the quantity of heat generated by completely combusting a specified quantity of hydrogen minus the heat of vaporization of the water in the combustion product. According to Chatterjee et al. (2014), the lower heating value of hydrogen is about 120 MJ/kg, which is much higher than those of CNG and gasoline. Hence, hydrogen will generate more energy per unit mass of hydrogen during its combustion than per unit mass of CNG and gasoline.

2.4.9 Octane Number

Octane numbers are a measure of the ability of a fuel to cause an engine to knock during operation. Knocking occurs when a secondary detonation fuel occurs inside an engine that leads to a temperature increase over and above the autoignition of the fuel, thereby causing the engine to knock. The higher the octane number of a fuel, the lower is its propensity to result into this unwanted combustion phenomenon.

2.4.10 Flash Point

Flash point is the temperature at which a fuel generates enough vapor to result into a flame when mixed with air and in the presence of an ignition source (Astbury, 2008; Hord, 1978). As a matter of fact, hydrogen has the lowest flash point among a wide range of common fuels, according to Mazloomi and Gomes (2012). Hydrogen-fired combustion devices are, therefore, expected to be much simpler to start and ignite than those fired with other fuels (Verhelst and Wallmer, 2009).

Other combustion properties worthy of mention include diffusivity and density which are higher and lower, respectively, than those of natural gas and gasoline. The high hydrogen diffusivity enables it to disperse thereby facilitating the formation of a uniform mixture with air, thereby enhancing its combustion. The very low density of hydrogen, however, presents it with two problems when utilized in an internal combustion engine – storage of a very large volume of hydrogen is required for an adequate driving range and reduced engine power output because of the energy density of hydrogen–air mixture.

In addition to the impact of hydrogen properties on its combustion behavior, impact of some hydrogen properties can be felt on its storage ability in underground reservoir. For example, it has been reported that diffusivity, along with some other properties (such as wettability, adsorption, interfacial tension), can significantly affect rock–fluid interactions during hydrogen storage, and consequently impact the safety of the storage process (Perera, 2023).

2.5 Thermal Properties

The thermal properties that are briefly discussed here are heat of vaporization, heat capacity, and thermal conductivity.

2.5.1 Heat of Vaporization

From thermodynamic viewpoint, the heat of vaporization (also referred to as heat of evaporation) is defined as the energy (enthalpy) that is required to convert a liquid substance to a gas. Although a constant heat of vaporization can be assumed for small temperature ranges, the heat of vaporization generally varies inversely with temperature. That is, it decreases with increasing temperature up to a certain point called the critical temperature where it completely decreases to zero. The liquid and vapor phases are not distinguishable above the critical temperature, at which point the substance is referred to as a supercritical fluid.

About 904 J (216 calories) of energy is required to transform a mole of liquid hydrogen to vapor, and it becomes a supercritical fluid at temperatures beyond –240°C and a pressure of 1.3 MPa. To put it in perspective, it takes about 40 times the amount of energy required to vaporize hydrogen to vaporize the same quantity of water, which requires about 40,000 J of energy per mole (Jolly, 2024).

2.5.2 Molar Heat Capacity

The molar heat capacity is an intensive property of a fuel, an intrinsic characteristic that does not depend on the size or shape of the amount in consideration (Mills et al., 1993). It is simply defined as the heat capacity of a chemical substance which is the amount of energy required to raise the temperature of a fuel by one unit per mole. It requires about 29 J of energy per role of hydrogen per Kelvin. Molar heat capacity, as a property, is most relevant in chemistry, when the quantity is specified on a molar basis rather than on mass or volume basis. The molar heat capacity generally increases with the molar mass, often varying with temperature, pressure, and state of matter.

2.5.3 Thermal Conductivity

Thermal conductivity is defined as the measure of a body's ability to conduct heat which occurs from energy transportation due to random movement of molecules across a temperature gradient. Hydrogen tends to have a higher thermal conductivity relative to other gases (about 0.2 W/(m-K)), which allows hydrogen to be used as the characteristic gas for gas analysis in detectors for gas chromatography, among other uses.

References

Abdel-Aal, H.K., Sadik, M., Bassyouni, M., and Shalabi, M. (2005). A New Approach to Utilize Hydrogen as a Safe Fuel. *Int. J. Hydrogen Energy*, 30, 1511–1514.

Al-Khalili, J. (2010). Discovering the Elements. Chemistry: A Volatile History. A Presentation in BBC. BBC Four.

Astbury, G.R. (2008). A Review of the Properties and Hazards of Some Alternative Fuels. *Process Safe Environ. Prot.*, 86, 397–414.

Balat, M. (2008). Potential Importance of Hydrogen as a Solution to Environmental and Transportation Problems. *Int. J. Hydrogen Energy*, 39, 4013–4029.

Bhadhesia, H.K.D.H. (2016). Prevention of Hydrogen Embrittlement in Steels. *ISIJ Int.*, 56, 24–36. Phase Transformations & Complex Properties Research Group, Cambridge University.

Broad, W.J. (1991). Breakthrough in Nuclear Fusion Offers Hope for Power of Future. *The New York Times*, November 1 Issue.

Chatterjee, A., Dutta, S., and Mandal, B.K. (2014). Combustion Performance and Emission Characteristics of Hydrogen as an Internal Combustion Engine Fuel. *J. Aeronautical Automot. Eng.*, 1(1), 1–6.

Clayton, D.D. (2003). *Handbook of Isotopes in the Cosmos: Hydrogen to Gallium*. Cambridge, England: Cambridge University Press. ISBN 978-0-521-82381-4.

Cordero, B., Gomez, V., and Platero-Plats, A.E. (2008). Covalent Radii Revisited. *Dalton Transactions*, 21(21), 2832–2838.

Dana, J.D. and Hurlbut, C.S. (1959). *Dana's Manual of Mineralogy* (17th ed.), pp. 78–89. New York: Chapman Hall.

Dunn, S. (2002). Hydrogen Future: Toward a Sustainable Energy System. *Int. J. Hydrogen Energy*, 3(8), 16–248.

Goldberg, D.E. (1988). *Three Thousand Solved Problems in Chemistry* (1st ed.), Section 17.43, p. 321. New York, NY: McGraw-Hill. ISBN 0-07-023684-4

Grolier, Inc. (1991). Hydrogen. *Grolier Encyclop. Know.*, 9, 401–402.

Gurov, Y.B., Aleshkin, D.V., Behr, M.N., Lapushkin, S.V., Morokhov, P.V., Pechkurov, V.A., Poroshin, N.O., Sandukovsky, V.G., Tel'kushev, M.V., Chernyshev, B.A., and Tschurenkova, T.D. (2004). Spectroscopy of Superheavy Hydrogen Isotopes in Stopped-Pion Absorption by Nuclei. *Phys. Atomic Nuclei*, 68(3), 49–97.

Haiyang, Y. (2009). Discrete Dislocation Plasticity HELPs Understand Hydrogen Effects in Bcc Materials. *J. Mech. Phys. Solids*, 123, 41–60.

Holte, A.E., Houck, M.A., and Collie, N.L. (2004). Potential Role of Parasitism in the Evolution of Mutualism in Astigmatid Mites. *Exp. Appl. Acarology*, 25(2), 97–107. doi:10.1023/A:1010655610575. PMID 11513367. S2CID 13159020.

Hord, J. (1978). Is Hydrogen a Safe Fuel? *Int. J. Hydrogen Energy*, 3, 57–76.

International Union of Pure and Applied Chemistry (IUPAC). (1997). *Electronegativity. Compendium of Chemical Terminology* (2nd ed.). Beverly Hill, CA: The Gold Book.

Jolly, W.L. (2024). Hydrogen. *Encyclopedia Britannica.* www.britannica.com/science/hydrogen.

Kendall, C. and Caldwell, E. (1998). Chapter 2: Fundamentals of Isotope Geochemistry. In *Isotope Tracers in Catchment Hydrology*, Kendall, C., and McDonnell, J. J. (Eds.), pp. 51–86. Reston, VA: US Geological Survey. doi:10.1016/B978-0-444-81546-0.50009-4.

Klerke, A., Christensen, C.H., Nørskov, J.K., et al. (2008). Ammonia for Hydrogen Storage: Challenges and Opportunities. *J. Mater. Chem.*, 18, 2304.

Korsheninnikov, A., Nikolskii, E., Kuzmin, E., Ozawa, A., Morimoto, K., Tokanai, F., Kanungo, R., Tanihata, I., et al. (2003). Experimental Evidence for the Existence of ^{7}H and for a Specific Structure of ^{8}He. *Phys. Rev. Lett.*, 90(8), 082501.

Kothari, R., Buddhi, D., and Sawhney, R.L. (2004). Sources and Technologies for Hydrogen Production: A Review. *Int. J. Global Energy Issues*, 21, (1/2), 154–178.

Laurendeau, N.M. and Glassman, I. (1971). Ignition Temperatures of Metals in Oxygen Atmospheres. *Combust. Sci. Tech.*, 3(2), 77–82.

Lewis, A.C. (2021). Optimizing Air Quality Co-Benefits in a Hydrogen Economy: A Case for Hydrogen-Specific Standards for NOx Emissions. *Environ. Sci. Atmos.*, 1(5), 201–207.

Lide, D.R., Ed. (2006). *CRC Handbook of Chemistry and Physics* (87th ed.). Boca Raton, FL: CRC Press. ISBN 0-8493-0487-3.

Mazloomi, K. and Gomes, C. (2012). Hydrogen as an Energy Carrier: Prospects and Challenges. *Renewable Sustain. Energy Rev.*, 16, 3025–3033.

Miessler, G.L. and Tarr, D.A. (1999). *Inorganic Chemistry* (2nd ed.), p. 41. Sadle River, NJ: Prentice Hall. ISBN 0-13-841891-8.

Mills, I., Cvita, T., Homann, K., Kallay, N., and Kuchitsu, K. (1993). Quantities, Units and Symbols. In International Union of Pure and Applied Chemistry (IUPAC), p. 7. Compendium of Chemical Terminology. Electronegativity (2nd ed). Oxford, England: Blackwell Sciences.

Momirlan, M. and Veziroglu, T.N. (2005). The Properties of Hydrogen as Fuel Tomorrow in Sustainable Energy System for a Cleaner Planet. *Int. J. Hydrogen Energy*, 30, 795–802.

Nave, C.R. (2006). *Deuterium-Tritium Fusion..* Atlanta, GA: HyperPhysics, Georgia State University.

Patnaik, P. (2007). *A Comprehensive Guide to the Hazardous Properties of Chemical Substances.* Hoboken, NJ: Wiley-Inter-science. p. 402. ISBN 978-0-471-71458-3.

Perera, M.S.A. (2023). A Review of Underground Hydrogen Storage in Depleted Gas Reservoirs: Insights into Various Rock–Fluid Interactions Mechanisms and their Impact on the Process Integrity. *Fuel*, 334(126677), 1–14.

Quaching, V. (2005). *Understanding Renewable Energy Systems.* UK: Carl-Hanser Ver-lag GmbH and Co. KIG.

Robertson, I.M. Sofronis, P., Nagao, A., Martin, M.L., Wang, S., Gross, D.W., and Nygren, K.E. (2015). Hydrogen Embrittlement Understood. *Metallurg. Mat. Transact. A*, 46A(6), 2323–2341.

Rogers, H.C. (1999). Hydrogen Embrittlement of Metals. *Science*, 159(3819), 1057–1064.

Schefer, E.W., Kulatilaka, W.D., Patterson, B.D., and Settersten, T.B. (2009). Visible Emission of Hydrogen Flames. *Combustion Flame*, 156(6), 1234–1241.

Stojic, D.L., Miljamic, S.S., Grozdic, T.D., Golobocamin, D.D., Sovili, S.P., and Jaksic, M.M. (1994). D/H Isotope Separation Efficiency in Water Electrolysis: Improvement by In Situ Activation at Different Temperatures. *Int. J. Hydrogen Energy*, 19, 587–590.

Stwertka, A. (1996). *A Guide to the Elements*, pp. 16–21. Oxford, England: Oxford University Press. ISBN 978-0-19-508083-4.

Theodore, L., Dupont, R., and Ganesan, K., Eds. (1999). *Pollution Prevention: The Waste Management Approach to the 21st Century*. Boca Raton, FL: CRC Press. 27, p. 15. ISBN 1-56670-495-2.

Urey, H.C., Brickwedde, F.G., and Murphy, G.M. (1933). Names for the Hydrogen Isotopes. *Science*, 78(2035), 602–603.

Verhelst, S. and Wallmer, T. (2009). Hydrogen-Fueled Internal Combustion Engine. *Progress Energy Combust Sci.*, 35, 490–527.

3

Hydrogen Production from Water

3.1 Introduction

As mentioned in Chapter 1, hydrogen can be produced from a variety of sources including fossil fuels, biomass/renewables, water, and/or a mixture of some or all the sources. Production of hydrogen via electrolysis has potential in contributing towards clean hydrogen production and addressing the greenhouse issue. Electrolysis-derived hydrogen can be produced with zero greenhouse gas emissions, depending on the source of energy and the technology used in generating the electricity needed for the electrolysis system.

Industrial production of hydrogen is currently mostly via steam reforming of primarily natural gas as feedstock. However, this method is not cost-effective enough because it requires high heat input and is less environmentally friendly from carbon dioxide (CO_2) emission viewpoint (Bespalko and Mizeraczyk, 2022a, 2022b). More environmentally friendly technologies (such as electrolysis, photobiological processes, and thermochemical processes) can be used to produce hydrogen. While some of these alternative methods are mature (e.g., electrolysis), the alternative methods need further development to be cost-effective with increased energy yield (Rosen and Scott, 1998). According to Mizeraczyk and Jasiński (2016) and as reported by Bespalko and Mizeraczyk (2022a,2022b) energy yield (grams of H_2 produced per kWh) of hydrogen production, which is the ratio of mass of hydrogen produced to the actual energy consumed in the process, determines the process competitiveness. In addition to hydrogen, several hydrogen-based products (such as ammonia, methanol, Fischer–Tropsch liquid fuels) can be produced from some of the sources (particularly, natural gas, coal, and biomass). While hydrogen production from natural gas and coal and from biomass will be dealt with in chapters 4 and 5, respectively, this chapter explores hydrogen production through water electrolysis.

Electrolysis is a promising procedure for carbon-free hydrogen production from water, and from some other renewable energy sources, such as nuclear. The history of hydrogen production via electrolysis dates to the early 19th century, when William Nicholson and Anthony Carlisle, in their experiment, observed the appearance of two types of gas bubbles (hydrogen and oxygen) when ends of two wires were connected to either side of a voltaic pile and the other ends of the wires were placed in a water-filled tube (Fabbri and Schmidt, 2018).

Water electrolysis is an electrochemical process that splits water into hydrogen and oxygen. While hydrogen produced from water and electricity (electric hydrogen) was

DOI: 10.1201/9781003348283-3

a major source of industrial hydrogen during the 1920s–1960s (IEA, 2019), hydrogen via water electrolysis currently accounts for only a small portion of the hydrogen production mix. According to the 2019 IEA report on The Future of Hydrogen, less than 0.1 percent of dedicated hydrogen (which is mainly used for products that require high-purity hydrogen, such as in electronics and polysilicon) production worldwide is from electrolysis (IEA, 2019).

While hydrogen production from water electrolysis appears to be attractive from a low carbon footprint perspective, this advantage can be negated if the electricity required for the electrolysis is derived from carbon-laden fossil fuels. As such, the source, efficiency, and associated emissions resulting from generating the required electricity for the electrolytic process are important factors that must be considered when analyzing and evaluating the viability of hydrogen production via electrolysis. However, with declining costs of renewable electricity, particularly from solar photovoltaic (PV) and wind, interest is growing in electrolytic hydrogen (IEA, 2019). The efficiency of electrolysis systems, which stands between 60 and 80 percent (IEA, 2019), is higher than that of electric power grid, not to mention the huge the amount of fuel (which consequently will lead to increased emission of greenhouse gases).

While electrolysis finds its application in many industrial processes (such as Hall–Heroult process for producing aluminum, electroplating, metal purification, chemical production, electrometallurgy, electrochemical machining, and rust removal), this chapter of the book focuses on fundamentals of hydrogen production via electrolysis technological pathway, including conventional electrolysis and unconventional electrolysis (plasma electrolysis, photo-electrolysis, and thermo-electrolysis).

3.2 Fundamentals

As shown in Figure 3.1, electrolysis is a process by which a direct electric current is passed through two electrodes (anode and cathode) in an electrolyte (water), thereby producing chemical reactions at the electrodes and, consequently, decomposing the material into its components (Kalamaras and Efstathiou, 2013). In other words, it is simply a process in which ionic substances are decomposed into simple substances by passing an electric current through them. In the case of hydrogen production using water electrolysis, water is split or decomposed into its two elements (hydrogen and oxygen). The reactions, occurring at the cathode and anode in the electrolyzer, are represented by equations [3.1] and [3.2], respectively and the overall reaction is represented by Equation [3.3] (Kalamaras and Efstathiou, 2013):

$$2H_2O\ (l) + 2e^- \rightarrow H_2\ (g) + 2OH^- \quad [3.1]$$

$$4OH^-\ (aq) \rightarrow O_2\ (g) + 2H_2O\ (l) + 4e^- \quad [3.2]$$

$$2H_2O\ (l) \rightarrow 2H_2\ (g) + O_2\ (g) \quad [3.3]$$

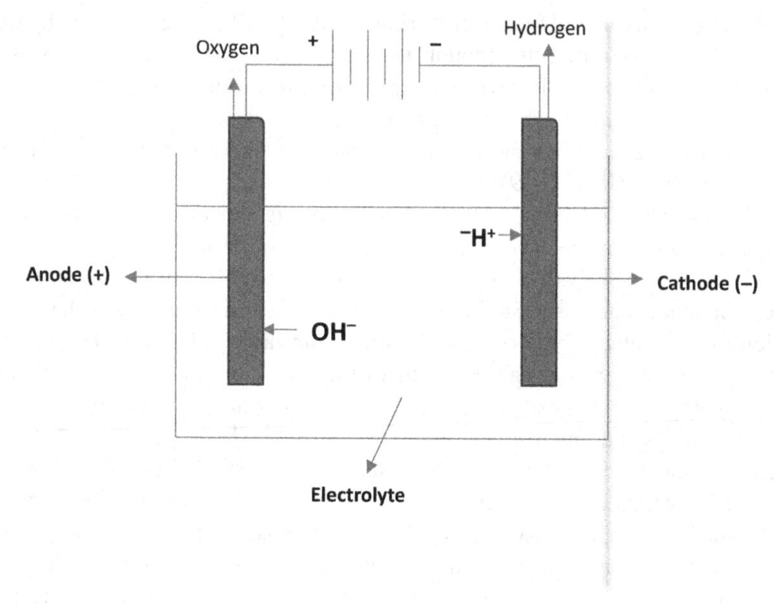

FIGURE 3.1
Schematic diagram of a simple water electrolysis system.

Electrolyzers can range in size from small, such as the ones in appliance-size equipment suitable for small-scale distributed hydrogen production, to large-scale, central production facilities that could be coupled directly to renewable or other non-greenhouse gas emitting electric power generation systems.

The major components of an electrolysis system are electrolyte, electrodes (anode and cathode), and an external power source. An electrolyte is a medium containing ion that is electrically conducting through the movement of those ions, but not conducting electrons (Enderby and Neilson, 1998; Petrovic, 2020; Winie et al., 2020). This includes most soluble salts, acids, and bases dissolved in a polar solvent, such as water. In nature, the atoms of electrolytes are closely bound together, which becomes weaker when dissolved and the molecules of electrolyte split into cations (positive ions) and anions (negative ions) moving freely in the solution. When the electrodes are immersed into the electrolyte and are connected to the direct current (DC) power source, the cations and the anions moving freely in the solution are attracted by the cathode and the anode, respectively. A partition (such as an ion-exchange membrane or a salt bridge) can be used (as an option) to keep the products from diffusing to the vicinity of the opposite electrode. However, it is worth noting that in some hydrogen production electrolysis process technologies, such as plasma electrolysis, metallic electrodes do not have to be immersed in water – as it is with traditional water electrolysis – to produce hydrogen, as corroborated by the study conducted by Chaffin et al. (2006). Instead, the electrodes can interact with water through atmospheric pressure plasmas located between the electrode tips and the water surface (Chaffin et al., 2006).

An electrode is an electrical conductor used to contact a non-metallic part of a circuit (e.g., a semiconductor, an electrolyte, a vacuum, or air). Electrodes are critical parts of batteries that can be made of a variety of materials depending on the type of battery. Platinum or any other inert metal is usually the material of choice for the cathode when producing hydrogen for storage. However, both electrodes could be made of inert metals for on-site hydrogen utilization, such as in combustion systems, where oxygen is required.

As noted earlier, conventional water electrolysis is a simple process by which hydrogen can be produced. It simply involves the basic electrochemical principle of dissociating water into hydrogen and oxygen when electrical energy is passed through water (as the electrolyte) between two electrodes (anode and cathode), as represented by Equation [3.3], according to Chaffin et al. (2006). During this process, gaseous oxygen and gaseous hydrogen form at the anode and the cathode, respectively (see Equation [3.1] and Equation [3.2]).

$$2H_2O\ (l) \rightarrow 2H_2\ (g) + O_2\ (g) \qquad [3.3]$$

This was the traditional principle formulated by Michael Faraday in the 19th century (Chaffin et al., 2006). As mentioned earlier, while Equation [3.3] represents the overall complete electrolysis process that occurs in the water electrolyzer, the stoichiometric and electrochemical reactions are as follows, according to Chaffin et al. (2006):

$$H_2O + E \rightarrow H^+ + OH^- \qquad [3.4]$$

$$H^+ + e^- \rightarrow H \qquad [3.5]$$

$$H + H \rightarrow H_2 \qquad [3.6]$$

where E is energy, H^+ is a positive proton, OH– is hydroxyl ion, H is hydrogen atom, and H_2 is a hydrogen molecule. In summary, in the electrolysis process, as shown in Equation [3.3], two moles of hydrogen gas and one mole of oxygen gas are produced from two moles of water. The hydrogen generated must diffuse out from large areas and through media of low transport resistance to promote high production and collection efficiency (Chaffin et al., 2006). Therefore, the system must be designed to ensure extraction of the produced gas bubbles generated, as demonstrated by Olthuis et al. (1998).

Electrolyzers are similar in design and nature to fuel cells. Different electrolyzers function in different ways, mainly determined by the type of electrolyte material involved and the ionic species conducted. The different electrolyzers include polymer electrolyte membrane electrolyzer, also known as proton exchange membrane (PEM), alkaline electrolyzer (AE), and solid oxide electrolyzers (SOE). While the alkaline-based electrolyzer seems to be the most common electrolysis technology, the development of PEM and SOE has been far ahead of AE (Grigoriev et al., 2006), as reported by Kalamaras and Efstathiou (2013). PEM electrolyzers tend to be free of corrosion issues and are more efficient than AE, and they are relatively more expensive than AE, while SOE technology is plagued with the problems of corrosion, thermal cycling, and seals (Kalamaras and Efstathiou, 2013).

In PEM electrolyzer, the electrons flow through an external circuit and the hydrogen ions selectively moves across the PEM to the cathode, where hydrogen ions combine with electrons from the external circuit to form hydrogen gas. The reactions at the anode and cathode are shown by equations [3.1] and [3.2], respectively (Kalamaras and Efstathiou, 2013).

According to Kalamaras and Efstathiou (2013), AEs are the most developed, the least efficient, and have the lowest capital cost among the three technologies. The principle of AEs typically involves movement of hydroxide ions (hydroxyl (OH–)) from the cathode through the electrolyte to the anode. In this case, the hydrogen is generated on the cathode side. Commercial electrolyte is a liquid alkaline solution of sodium or potassium hydroxide as the electrolyte has been available for many years (Saksono et al., 2012, 2014, 2017; Kalamaras and Efstathiou, 2013). Newer laboratory-scale approaches using solid alkaline exchange membranes (AEM) as the electrolyte are promising (Saksono et al., 2014] U.S. Department of Energy, Office of Energy Efficiency and Renewable Energy (U.S. DOE/EERE, 2016).

A solid ceramic material is used as the electrolyte for SOEs. The electrolyte selectively conducts negatively charged oxygen ions at elevated temperatures. In this case, the mechanism for hydrogen generation, as described by U.S. DOE/EERE (2016) is as follows:

- Steam at the cathode combines with electrons from the external circuit to form hydrogen gas and negatively charged oxygen ions.
- The oxygen ions pass through the solid ceramic membrane and react at the anode to form oxygen gas and generate electrons for the external circuit.

SOE has a drawback of requiring high operating temperatures (~800–1000°C) for the solid oxide membranes to function properly (U.S. DOE/EERE, 2016). As reported by IEA in their 2019 report, this is much higher than that of PEM electrolyzers, which operate at 70–90°C, and commercial alkaline electrolyzers, which typically operate at less than 100°C. Laboratory-scale research efforts to lower the operating temperature for SOE is promising (U.S. DOE/EERE, 2016).

The theoretical maximum efficiency is higher than 85 percent (Thomassen, 2018; Kruse et al., 2020). Typically, conventional water electrolysis for hydrogen production is reported to have efficiency of only about 30 percent (Rosen and Scott, 1998). This is significantly lower than that of SMR, which is about 80 percent (Rosen and Scott, 1998). A number of research studies (Rosen and Scott, 1998; Seehra et al., 2008, 2009; Seehra and Bollineni, 2009; Ranganathan, 2007; Bollineni, 2008; Akkineni, 2011; Coughlin and Farooque, 1980) have been successfully conducted to improve the efficiency of hydrogen production via electrochemical process. The results of these studies, which have been summarized by Rosen and Scott (1998), have shown that using carbon-assisted water electrolysis in which carbon is added to the anode can enhance efficiency and that hydrogen evolution could occur at a much lower applied voltages than required for conventional water electrolysis. Equation [3.7] represents the net reaction when carbon is added to the electrolyte (Rosen and Scott, 1998).

$$C\,(s) + 2H_2O \rightarrow CO_2\,(g) + 2H_2\,(g) \qquad [3.7]$$

resulting in evolution of CO_2 at the anode and pure hydrogen at the cathode. The reaction depicted by Equation [3.7] can then be claimed to be an electrochemical gasification of carbon in water occurring at room temperature with the concurrent evolution of pure hydrogen produced at the cathode.

3.3 Technologies

3.3.1 Conventional

As mentioned in Section 3.2, the main electrolysis technologies include: (1) alkaline electrolysis, (2) proton exchange membrane (PEM) electrolysis, and (3) solid oxide electrolysis cells (SOECs). There are two basic types of designs – unipolar and bipolar designs – in existence for each of the three electrolysis technologies (NREL, 2009). A unipolar electrolyzer essentially looks like a tank with its electrodes connected in parallel. The unipolar electrolyzer design is a high-current, low-voltage system with a single bus bar connecting all the anodes and all the cathodes connected by the other bar. A membrane is placed between each cathode and anode to ensure that the hydrogen and oxygen are separated as they are produced, at the same time allowing the transfer of ions (NREL, 2009). The bipolar AE functions like a filter press. Bipolar electrolyzer cells are connected in series, where hydrogen is produced on one side of each cell and oxygen on the other side. It is a high-voltage, lower-current unit. There is a membrane that separates the electrodes. Most commercial AE systems use the bipolar design (NREL, 2009).

3.3.1.1 Alkaline Electrolysis

Alkaline electrolysis is a mature and commercial technology that has been used for hydrogen production in the fertilizer and chlorine industries since the 1920s (IEA, 2019). AE operates at a temperature range of 60–80°C, a pressure range of 1–30 bars, and over a range of load varying from a minimum of 10 percent to full design capacity (IEA, 2019). According to IEA, several bigger capacities (of up to 165 megawatts electrical (MWe)) were built in the last century in different countries (such as Canada, Egypt, India, Norway, and Zimbabwe) with large hydropower resources.

However, almost all these AE plants have been replaced with natural gas and steam methane reforming.

AEs use an aqueous solution of potassium hydroxide (KOH) which allows it to take advantage of KOH's high conductivity, as well as the fact that the oxygen evolution reaction occurs with a low energy loss when KOH is used (NREL, 2009). AEs do not require precious metals as they typically use nickel electrodes. Also, AEs do not use precious materials, they hereby have relatively low capital costs compared with other electrolyzer technologies (NREL, 2009).

AEs have found their use in a few applications such as space, military, transportation, and electric-generation backup. Their advantages include wider range of stable materials, which allows the use of lower-cost materials, low operating temperature, and quick start-up (U.S. DOE/EERE, 2016). They are, however constrained by (NREL, 2009; U.S. DOE/EERE, 2016):

- Sensitivity to CO_2 that may be present in fuel and air
- Requirement for the management of the aqueous electrolyte
- Electrolyte conductivity

The U.S. DOE/EERE (2016) report indicated the existence of some commercial/near commercial AEs in some parts of the world (the United States, Norway, and Switzerland). According to the U.S. DOE/EERE (2016) study report, the US AE plant has a unipolar design with a production capacity of up to 10 kg/day of hydrogen, while the plants in Norway and Switzerland are both bipolar and are both at much higher hydrogen production capacity (1000 and 1500 kg/day for the plant in Norway and Switzerland, respectively) than the one in the United States (U.S. DOE/EERE, 2016).

3.3.1.2 Proton Exchange Membrane (PEM) Electrolysis

General Electric introduced PEM electrolyzer systems in the 1960s to ameliorate and solve some of the operational problems posed by AEs (IEA, 2019). Pure water was used as an electrolyte solution for the PEM system instead of potassium hydroxide (KOH) being used for AE. PEM electrolyzer systems are relatively small, rendering them potentially more amenable than AE for urban use (IEA, 2019). PEM can operate at a wide range of load varying from as low as 0 to 160 percent of design capacity, compared with a relatively narrower range for alkaline electrolysis and SOEC systems, which are 10–110 and 20–100 percent, respectively (IEA, 2019). They also operate at temperature range of 50–80°C and a pressure range of 30–80 bars. PEM systems are, however, faced with some drawbacks, which include high cost and short life (IEA, 2019). They are expensive because of their requirements for expensive electrode catalysts (such as platinum, iridium) and membrane materials for electrode. According to the IEA (2019) report, their overall costs are higher than those of AEs and they are less widely deployed.

PEM can be used for backup power generation, portable power generation, and distributed power generation, transportation, and in specialty vehicles. While it requires expensive catalysts and is sensitive to fuel impurities, it has the following advantages (U.S. DOE/EERE, 2016):

- Low corrosion and electrolyte management problems because of the solid electrolyte used
- Low operating temperature, thereby low costs, and
- Quick start-up

The U.S. DOE/EERE (2016) study report indicated the existence of some PEM commercial/near commercial electrolyzers of bipolar design in the United States.

3.3.1.3 Solid Oxide Electrolysis Cell

SOECs electrolysis technology is yet to be commercialized and is the least developed, operating at a relatively low pressure (1 bar) and a relatively higher temperature range (680–1000°C), compared with AE and PEM, as reported by IEA (2019). However, effort by individual companies is ongoing to bring them to market (IEA, 2019). Ceramic electrolyte, which is relatively inexpensive, is common with SOECs. Steam, which must be generated from an external source, usually serves as the electrolyte for SOEs. The heat source could be from a wide range of sources, including fossil fuel, nuclear, solar, geothermal, or integrated synthesis processes (such as Fischer–Tropsch synthesis and methanation).

Unlike AEs and PEM electrolyzers, SOEC electrolyzer can be designed and operated in reverse mode, that is, as a fuel cell, whereby hydrogen can be converted back into electricity, thereby serving as a hydrogen storage facility (IEA, 2019). SOEC electrolyzer has also been used for combined-steam and carbon dioxide electrolysis, producing syngas (a gas mixture of carbon monoxide and hydrogen) for subsequent conversion to a synthetic fuel. One key issue with designing and developing SOEC electrolyzers is addressing the material degradation that can result from the high operating temperatures (IEA, 2019).

According to IEA (2019), there has been an increase in new electrolysis installations over the last decade aimed at producing hydrogen from water, with PEM technology making significant progress in market penetration. Geographically most of the projects are in Europe, although projects have also been started or announced in Australia, China, and the Americas, as enumerated by the IEA (2019) report. The average unit size of these electrolyzer additions has increased over the years from 0.1 MWe during 2000–2009 to 1.0 MWe during 2015–2019, indicating a shift from small pilot and demonstration projects to commercial-scale applications (IEA, 2019).

SOEs are commonly used in electric power generation plants, distributed electric plants, and auxiliary equipment. Their advantages include (U.S. DOE/EREE, 2016):

- High efficiency
- Fuel flexibility

- Solid electrolyte
- Suitable for combined heat and power (CHP) systems
- Hybrid/gas turbine cycle

According to a report by the U.S. DOE/EREE (2016), there are, however some drawbacks of SOEC, such as:

- High-temperature corrosion and breakdown of cell components
- Long start-up time
- Limited number of shutdowns

3.3.2 Unconventional Electrolysis

As mentioned earlier in the chapter, the efficiency of conventional electrolysis process for hydrogen tends to be relatively low compared with fossil-fuel-derived hydrogen (Rosen and Scott, 1998). As such, more efficient electrolysis processes need to be developed for electrolysis-based technologies to play a major role in the hydrogen economy. A description of some new unconventional technologies based on electrolysis (plasma electrolysis, photo-electrolysis, and thermal electrolysis) will be the focus of this section of the chapter.

3.3.2.1 Plasma Electrolysis

Another type of electrolysis that can be used for hydrogen production is the plasma-driven electrolysis (PDE) (also known as the contact glow-discharge electrolysis (CGDE), as reported by Bespalko and Mizeraczyk (2022a, 2022b)). Plasma-driven electrolysis is an unconventional electrochemical process, which involves the formation of electric plasma by glow discharges from direct or pulsed currents in gas-vapor envelopes. These envelopes are produced in the vicinity of discharge electrodes submerged in electrolytic solutions (Bespalko and Mizeraczyk, 2022b). PDE is, therefore, basically an electrolysis process conducted under high voltage to produce electrical spark, resulting in the formation of plasma in an electrolyte (Bespalko and Mizeraczyk, 2022a). Several researchers, including Saksono et al. (2014, 2017) ~~and,~~ have conducted various studies on the use of plasma for producing hydrogen. More detailed description of the fundamental principles of PDE process technology can be found in an overview paper by Bespalko and Mizeraczyk (2022a and elsewhere (Gupta et al., 2016; Mota-Lima et al., 2019; Bruggeman et al., 2021; Mizuno et al., 2002, 2004, 2005). Studies conducted by Mizuno and co-workers (2002, 2004, 2005) have shown that PDE efficiency for hydrogen production is far higher than that in Faradaic electrolysis.

This simply infers that higher yield (which refers to the ratio of the product to the mole electrons consumed in a chemical reaction) is achieved by PDE system compared with those produced via Faraday system. Also, PDE has the capabilities to produce other valuable products (such as chemical compounds), which are lacking with Faradaic electrolysis (Mills et al., 2002).

While it is known (as described earlier) that hydrogen can be produced from water via electrolysis, a much higher rate of production can be achieved at higher flow rate of the electrolyte (Mat et al., 2004) provided some other voltage requirements (such as reversible potential, over voltage on electrodes, and ohmic loss in aqueous solution) are in place (Nagai et al., 2003; Chaffin et al., 2006), which are only achievable at high temperature and pressure. A review of the basic theory of movement of charges resulting from an electric field in an ionized plasma region is given by Chaffin et al. (2006). Understanding of this basic theory is essential to identify the key parameters (mobility, electric field, and drift velocity due to the field) in hydrogen production via PDE technology.

In their experimental research study, Chaffin et al. (2006) revealed that PDE may provide a novel way of producing hydrogen at a sufficiently fast rate, which may be adequate for incorporation into cost-effective onsite hydrogen utilization systems. Chaffin et al.'s (2006) study further showed that the surface current that primarily flows near the water–plasma interface, rather than through the liquid bulk, is critical to the PDE process. The results of the study by Saksono et al. (2017) concur with the observation made by Chaffin et al. (2006), which suggest that hydrogen can be produced at a significantly higher rate using PDE than via the conventional Faraday electrolysis technique, as mentioned earlier.

3.3.2.2 Photo-Electrolysis

Photo-electrolysis is one of the promising, efficient, and cost-effective unconventional electrolysis methods of producing hydrogen, especially from renewable source despite it being presently at an experimental stage of development (Kalamaras and Efstathiou, 2013; Hamelinck and Faaij, 2002). As reported by Kalamaras and Efstathiou (2013), the system utilizes a semiconducting photoelectrode that absorbs solar energy and simultaneously generates the required voltage for the direct decomposition of water molecule into its components – oxygen and hydrogen. Hence, photo-electrolysis is essentially a process that uses a photoelectrochemical (PEC) light collection system for carrying out water electrolysis. The energy required to sustain the system can be generated when the solar-powered semiconductor photoelectrode is submerged in an aqueous electrolyte, according to Kalamaras and his co-worker (2013). However, the reaction depends on the solar intensity (which produces adequate current density) and on the type of semiconductor material.

The photoelectrode is made up of photovoltaic (semiconductor), catalytic, and protective layers, which can be modeled as independent components (Lindquist and Fell, 2009). Each photovoltaic layer (which is produced from light-absorbing semiconductor materials) influences the overall efficiency of the photoelectrochemical system (Kalamaras and Efstathiou, 2013). The light absorption capacity of the semiconductor material is directly proportional to the performance of the photoelectrode, according to Kalamaras and Efstathiou (2013). The required potential for splitting the water molecule is provided by semiconductors with wide bands (Acar and Dincer, 2018; Kalamaras and Efstathiou, 2013).

The performance of the photochemical electrolysis process (which requires suitable catalysts) is also claimed to be influenced by the catalytic layers of the photoelectrochemical cell (Kalamaras and Efstathiou, 2013).

3.3.2.3 Thermochemical Water Splitting

Another method by which water can be decomposed into its components to produce hydrogen is via thermochemical process, which was reported to have been developed during the 1970s and 1980s (Kalamaras and Efstathiou, 2013). Thermochemical water splitting (also known as thermolysis) is a process where water is thermally dissociated into hydrogen and oxygen (Steinfeld, 2005), with a potential to achieve overall efficiency of about 50 percent (Funk, 2001). The overall single-step thermal water decomposition reaction can be represented by Equation [3.8] (Kalamaras and Efstathiou, 2013):

$$H_2O + \text{Heat} \rightarrow H_2 + 0.5O_2 \qquad [3.8]$$

This process is, however, plagued with the need for an effective method (such as the use of semipermeable membranes with high-temperature materials based on ZrO_2) for separating hydrogen and oxygen to prevent the formation of an explosive mixture (Kalamaras and Efstathiou, 2013). This can be effectively achieved by using palladium membranes.

Although water will decompose at around 250°C, not many materials can remain stable at this temperature (Kalamaras and Efstathiou, 2013). Research efforts have been and are being successfully made to lower the decomposition temperature (Lewis et al., 2003). In addition to lowering the water dissociation temperature, the results of these research efforts have yielded a better understanding of the relationship between capital cost, thermodynamic losses, and process thermal efficiency, which may lead to reduced hydrogen production costs (Funk, 2001).

References

Acar, C. and Dincer, I. (2018). Hydrogen Production. *Compr. Energy Syst.*, 3, 1–40.

Akkineni, L.P. (2011). Hydrothermal Pretreatment of Biomass Samples for Producing Energy Efficient Hydrogen Electrochemically. M.Sc. Thesis, West Virginia University.

Bespalko, S. and Mizeraczyk, J. (2022a). Plasma Discharges in the Anodic and Cathodic Regimes of Plasma Driven Solution Electrolysis for Hydrogen Production. *Przegląd Elektrotechniczny*, 1, 124–127.

Bespalko, S. and Mizeraczyk, J. (2022b). Overview of Hydrogen Production by Plasma-Driven Solution Electrolysis. *Energies*, 15, 7508. https://doi.org/10.3390/en15207508

Bollineni, S. (2008). Hydrogen Production via Carbon- Assisted Water Electrolysis at Room Temperature: Effects of Catalysts and Carbon Type. M.Sc. Thesis, West Virginia University.

Bruggeman, P.J., Frontiera, R.R., Kortshagen, U.R., Kushner, M.J., Linic, S., Schatz, G.C., Andaraarachchi, H., Exarhos, S., Jones, L.O., Mueller, C.M., et al. (2021). Plasma-Driven Solution Electrolysis. *J. Appl. Phys.*, 129, 200902.

Chaffin, J.H., Bobbio, S.M., Inyang, H.I., and Kaanagbara, L. (2006). Hydrogen Production by Plasma Electrolysis. *J. Energy Engineering*, 132, 3.

Coughlin, R.W. and Farooque, M. (1980). Electrochemical Gasification of Coal: Simultaneous Production of Hydrogen and Carbon Dioxide by a Single Reaction Involving Coal, Water and Electrons. *Ind. Eng. Chem. Process Des. Dev.*, 19, 211.

Enderby, J.E. and Neilson, G.W. (~~1981~~ 1998). The Structure of Electrolyte Solutions. *Reports Prog. Phys.*, 44(6), 593–653.

Fabbri, E. and Schmidt, T.J. (2018). Oxygen Evolution Reaction—The Enigma in Water Electrolysis. *ACS Catalysis*, 8(10), 9765–9774. doi:10.1021/acscatal.8b02712.

Funk, J.E. (2001). Thermochemical Hydrogen Production: Past and Present. *Int. J. Hydrogen Energy*, 26(3), 185–190.

Grigoriev, S.A., Porembsky, V.I., and Fateev, V.N. (2006). Pure Hydrogen Production by PEM Electrolysis for Hydrogen Energy. *Int. Journal Hydrogen Energy*, 31(2), 171–175.

Gupta, S.K.S. and Singh, R. (2016). Cathodic Contact Glow Discharge Electrolysis: Its Origin and Non-Faradaic Chemical Effects. *Plasma Sources Sci. Technol.*, 26, 015005.

Hamelinck, C.N. and Faaij, A.P.C. (2002). Prospects for Production of Methanol and Hydrogen from Biomass. *J. Power Sources*, 111(1), 1–22.

IEA. (2019). The Future of Hydrogen – Seizing Today's Opportunities. Report Prepared by IEA for the G20, Japan.

Kalamaras, C.M. and Efstathiou, A.M. (2013). Hydrogen Production Technologies: Current State and Future Developments. Paper Presented at Proceedings of the Power Options for the Eastern Mediterranean Region Conference held 19-21 November 2012 in Limassol, Cyprus. Hindawi Publishing Corporation: Chichester, West Sussex, England.

Kruse, B., Grinna, S., and Buch, C. (2020). Hydrogen Status of Muligheter. Bellona. Archived from the European Hydrogen Week.

Lewis, M.A., Serban, M., and Basco, J.K. (2003). Hydrogen Production at >550°C Using a Low Temperature Thermochemical Cycle. In Proceedings of the Atoms for Prosperity: Updating.

Lindquist, E.S. and Fell, C. (2009). Fuels-Hydrogen Production: Photoelectrolysis. In *Encyclopedia of Electrochemical Power Sources*, Jurgen, G. (Ed.), pp. 369–383. Elsevier: Amsterdam, The Netherlands.

Mat, M.D., Aldas, K., and Ilegbusi, O.J. (2004). A Two-Phase Flow Model for Hydrogen Evolution in an Electrochemical Cell. *Int. J. Hydrogen Energy*, 29, 1015–1023.

Mills, R., Dayalan, E., Ray, P., Dhandapani, B., and He, J. (2002). Highly Stable Novel Inorganic Hydrides from Aqueous Electrolysis and Plasma Electrolysis. *Electrochim. Acta*, 47, 3909–3926.

Mizeraczyk, J. and Jasiński, M. (2016). Plasma Processing Methods for Hydrogen Production. *Eur. Phys. J. Appl. Phys.*, 75, 24702.

Mizuno, T., Akimoto, T., Azumi, K., Ohmori, T., Aoki, Y., and Takahashi, A. (2005). Hydrogen Evolution by Plasma Electrolysis in Aqueous Solution. *Jpn. J. Appl. Phys.*, 44, 396–401.

Mizuno, T., Akimoto, T., and Ohmori, T. (2002). Confirmation of Anomalous Hydrogen Generation by Plasma Electrolysis. In Proceedings of the 4th Meeting of Japan CF Research Society, Iwate, Japan, 17–18 October 2002.

Mizuno, T., Aoki, Y., Chung, D.Y., Sesftel, F., and Biberian, J.-P. (2004). Generation of Heat and Products During Plasma Electrolysis. In Proceedings of the 11th International Conference on Cold Fusion, pp. 161–177, Marseilles, France.

Mota-Lima, A., Nascimento, J.F.D., Chiavone-Filho, O., and Nascimento, C.A.O. (2019). Electrosynthesis via Plasma Electrochemistry: Generalist Dynamical Model to Explain Hydrogen Production Induced by a Discharge over Water. *J. Phys. Chem.*, 123, 21896–21912.

Nagai, N., Takeuchi, M., Kimura, T., and Oka, T. (2003). Existence of Optimum Space Between Electrodes on Hydrogen Production by Water Electrolysis. *Int. J. Hydrogen Energy*, 28, 35–41.

National Renewable Energy Laboratory (NREL) of the U.S. Department of Energy. (2009). Current State-of-the-Art Hydrogen Production Cost Estimate Using Water Electrolysis.

Olthuis, W., Volanschi, A., and Bergveld, P. (1998). Dynamic Surface Tension Measured with an Integrated Sensor–Actuator Using Electrolytically Generated Gas Bubbles. *Sens. Actuators B*, 49, 126–132.

Petrovic, S. (2020). *Battery Technology Crash Course: A Concise Introduction*. Springer Publishing: New York, NY. ISBN 978-3-030-57269-3. OCLC 1202758685.

Ranganathan, S. (2007). Carbon Assisted Electrolysis of Water to Produce Hydrogen at Room Temperature. M.Sc. Thesis, West Virginia University.

Rosen, M.A. and Scott, D.S. (1998). Comparative Efficiency Assessments for a Range of Hydrogen Production Processes. *Int. J. Hydrogen Energy*, 23, 653–659.

Saksono, N., Ariawan, B., and Bismo, S. (2012). Hydrogen Production System Using Non-Thermal Plasma Electrolysis in Glycerol-KOH Solution. *Int. J. Tech.*, 1, 8–15.

Saksono, N., Batunbara, T., and Bismo, S. (2017). Hydrogen Production by Plasma Electrolysis Reactor of KOH-Ethanol Solution. Proceedings of the 2nd International Conference on Chemical Engineering, IOP Conference Series: Material Science and Engineering 162, 012010, doi:10.1088/1757-899X/162//1/012010.

Saksono, N., Faishal, M., Merisa, B., and Setidjo, B. (2014). Hydrogen Production System with Plasma Electrolysis Method in Natrium Carbonate-Acetate Acid Solution. Proceeding the Regional on Chemical Engineering, Yogyakarta, Indonesia.

Seehra, M.S., Akkineni, L.P., Yalamanchi, M., Singh, V., and Poston, J. (2009). Structural Characteristics of Nanoparticles Produced by Hydrothermal Pretreatment of Cellulose and Their Applications for Electrochemical Hydrogen Generation. *Int. J. Hydrogen Energy*, 37, 9514–9523.

Seehra, M.S. and Bollineni, S. (2009). Nanocarbon Boosts Energy-Efficient Hydrogen Production in Carbon-Assisted Water Electrolysis. *Int. J. Hydrogen Energy*, 34, 6078–6084.

Seehra, M.S., Ranganathan, S., and Manivannan, A. (2008). Carbon-Assisted Water Electrolysis: An Energy-Efficient Process to Produce Pure H_2 at Room Temperature. *Appl. Phys. Lett., Erratum: Appl. Phys. Lett.*, 92, 239902, 044104(3).

Steinfeld, A. (2005). Solar Thermochemical Production of Hydrogen—A Review. *Solar Energy*, 78(5), 603–615.

Thomassen, M. (2018). Cost Reduction and Performance Increase of PEM Electrolyzers. fch.europa.eu. FCH JU. Archived from the European Hydrogen Week.

U.S. Department of Energy, Office of Energy Efficiency and Renewable Energy. (2016). Comparison of Fuel Cell Technologies.

Winie, T., Arof, A.K., and Thomas, S. (2020). *Polymer Electrolytes: Characterization Techniques and Energy Applications*. John Wiley & Sons: Hoboken, NJ. ISBN 978-3-527-34200-6.

4

Hydrogen Production from Fossil Fuels

4.1 Preamble

Hydrogen appears to be a very good candidate to contribute towards addressing climate issues. By virtue of the name, hydrocarbons are predominantly made up of carbon and hydrogen atoms (Zuttel et al., 2008; DeLuchi, 1989; Balat and Kirtay, 2010), from which hydrogen can be produced (Mazloomi and Gomes, 2012). The fundamental core of hydrogen value chain is its production, which has been growing rapidly (at about 8–10 percent per annum globally) for many years according to Kothari et al. (2004). Hydrogen can be manufactured from a wide range of sources, including fossil fuels (especially, natural gas and coal), biomass, and water using a variety of technologies and processes or pathways via chemical, thermal, and biological processes. According to Hassman and Kuhme (1993), fossil fuels account for over 90 percent of the hydrogen used by the refinery and ammonia production industry.

The various ways in which hydrogen can be produced include thermal (natural gas reforming, coal and biomass gasification, and pyrolysis), biological, electrolytic, photolytic, and electrochemical. Hydrogen production from water and biomass is discussed in chapters 3 and 5, respectively, whereas insights in the fundamentals and technologies of producing hydrogen from natural gas and coal is presented in this chapter.

The bulk (nearly about 50 percent) of the global demand of hydrogen produced today is by steam reforming of natural gas (Balat and Balat, 2009; Konieczny et al., 2008; Kalamaras and Efstathiou, 2013). Next to natural gas steam reforming is oil/naphtha reforming from refinery/chemical industrial off-gases, which accounts for about 30 percent of global hydrogen demand, while 18 percent and about 4 percent are from coal gasification and water electrolysis, respectively (Mazloomi and Gomes, 2012; Muradov and Veziroglu, 2005).

For the most part, the technology/process utilized for manufacturing hydrogen from each of the sources varies with the feedstock. For example, the main technologies for producing hydrogen from natural gas include steam methane reforming (SMR), autothermal reforming (ATR), partial oxidation (POX), pyrolysis, and plasma reforming (PR), while the technologies of interest for hydrogen production from coal include gasification and pyrolysis.

Different versions of these technologies have been developed by different companies over the years. Some of the companies that have developed gasification

DOI: 10.1201/9781003348283-4

technology include Lurgi, Shell, Texaco, and General Electric. This chapter presents the fundamentals of hydrogen production from fossil fuels (natural gas and coal).

4.2 Relevant Properties of Natural Gas and Coal

Natural gas and coal are carbonaceous materials that have chemical and physical properties that play important roles in their conversion to hydrogen or hydrogen-based products. These include the choice and design of the most appropriate conversion process or technology and product characterization (Probstein and Hicks, 1982). Given further is a brief description of some of the most relevant characteristics of natural gas and coal. More details can be found elsewhere (Probstein and Hicks, 1982; Ogunsola and Lam, 1993; Morgan et al., 1981).

4.2.1 Natural Gas

Natural gas is a naturally occurring mixture of hydrocarbon gases which consist mainly of methane along with various smaller amounts of other higher hydrocarbon gases. Small concentrations of some other gases (such as carbon dioxide, nitrogen, hydrogen sulfide, and helium) are also known to be usually present (Naturalgas.org, 2014).

Like other fossil fuels (oil and coal), many different theories have been propounded for the origin and formation of natural gas. It is formed from the remains of plants and animals and microorganisms that lived millions and millions of years ago. Natural gas is formed when layers of organic matter decompose under anaerobic conditions and are subjected to intense heat and pressure underground over millions of years (Kashtan et al., 2023).

Natural gas can be burned for heating, cooking, and electricity generation. It is also used to manufacture organic chemicals, including hydrogen. Concentration of the various components of natural gas varies with gas field and location (U.S. Department of Energy (DOE), Energy Information Administration (DOE/EIA), 2020). According to the information obtained from Naturalgas.org website, a typical natural gas is made up of the following components and concentrations – methane (70–90 percent), ethane + propane + butane (0–20 percent), CO_2 (0–8 percent), oxygen (0–0.2 percent), nitrogen (0–5 percent), hydrogen sulfide (0–5 percent), and trace amounts of other inert gases such as argon and helium. Hence, it is obvious that natural gas is predominantly made of methane, ranging from 87 to 98 mole percent (typically about 95 mole percent), followed by ethane, ranging from 1.5 to 9 mole percent (with typical value of about 4 mole percent). The other four hydrocarbon gases (propane, butane, pentane, and hexane) are when combined comprise less than 0.4 mole percent. Also of note is the presence of hydrogen, some corrosive acid gases (H_2S and CO_2), some diluent (helium and nitrogen) in some natural gas streams. As a matter of fact, some natural gas streams could be a commercial source for helium (Kidnay and Parrish, 2006). One of the important objectives of natural gas processing is to remove the corrosive and toxic gas H_2S and convert it to elemental sulfur. Other impurities that may be found in some natural gas streams are given elsewhere (Kidnay and Parrish, 2006).

Since methane (CH_4) is the largest component of natural gas, it is therefore appropriate that the conversion of natural gas can be well represented by methane conversion, which can be easily converted to hydrogen via thermal processes, such as SMR or steam natural gas-reforming (SNGR), ATR, and POX. It can also be pyrolyzed to produce hydrogen, as well as through plasma reforming (PR).

4.2.2 Coal

Coal, which contributes about 18 percent of the world hydrogen demand, can be converted to hydrogen through gasification and pyrolysis. Coals can be classified by two main parameters – type and rank – the most important of which is rank. Coal type is generally determined by the plant materials from which it was formed, or it is composed of, while rank denotes the extent of its maturation or age during the formation process (coalification). Coal is formed from plant, tree, and woody materials buried in great depth over hundreds of million years ago. The burial of the dead, decayed plant materials resulted in the chemical changes of the materials, thereby leading to the formation of coal of different rank at varying depth. There are four main coal ranks (lignite, subbituminous coal, bituminous coal, and anthracite), with lignite being the lowest range and the youngest coal, followed in increasing rank and age by subbituminous, bituminous, and anthracite.

Since coal originated in peat deposits and the dominant material was wood (Probstein and Hicks, 1982), the chemical characteristics of coal would be expected to be like those of peat and wood. The average value (obtained from Probstein and Hicks, 1982) of some key chemical elements (carbon, hydrogen, and oxygen) in wood, peat, and various coal ranks reveals the variation of these properties with coalification. The variability is graphically represented in Figure 4.1(a,b) to further illustrate changes in these elements as a function of coalification. It is evident from Figure 4.1(a,b) that peat, wood, and all the coals of different ranks contain the same elements, but with varying concentration. Peat and wood appear to contain more oxygen and hydrogen than coal (Figure 4.1(b)), while coal contains more carbon than wood and peat (Figure 4.1(a)). Also shown in Figure 4.1(a,b) is the variation in the chemical characteristics of coal with rank. It is worth noting that the values plotted in Figure 4.1 are only representative averages. It could be observed that the carbon content of the coal increases with rank (Figure 4.1(a)), while the oxygen and the hydrogen content decrease with increase in rank (Figure 4.1(b)). From the coal-to-hydrogen (CTH) standpoint, the carbon and hydrogen contents are critical as they are the key elements involved in the conversion process.

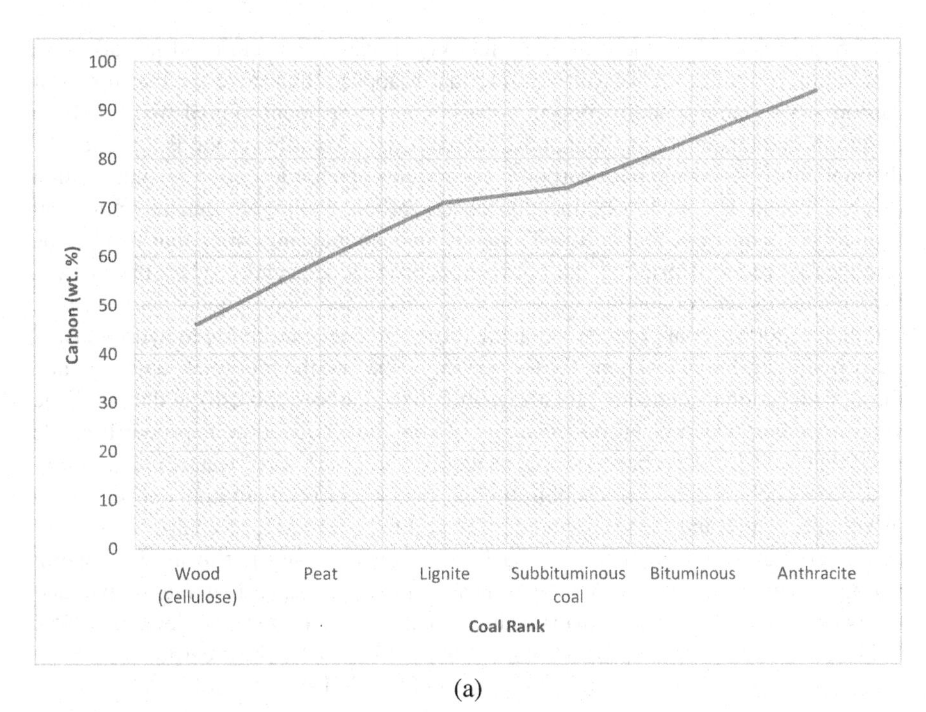

(a)

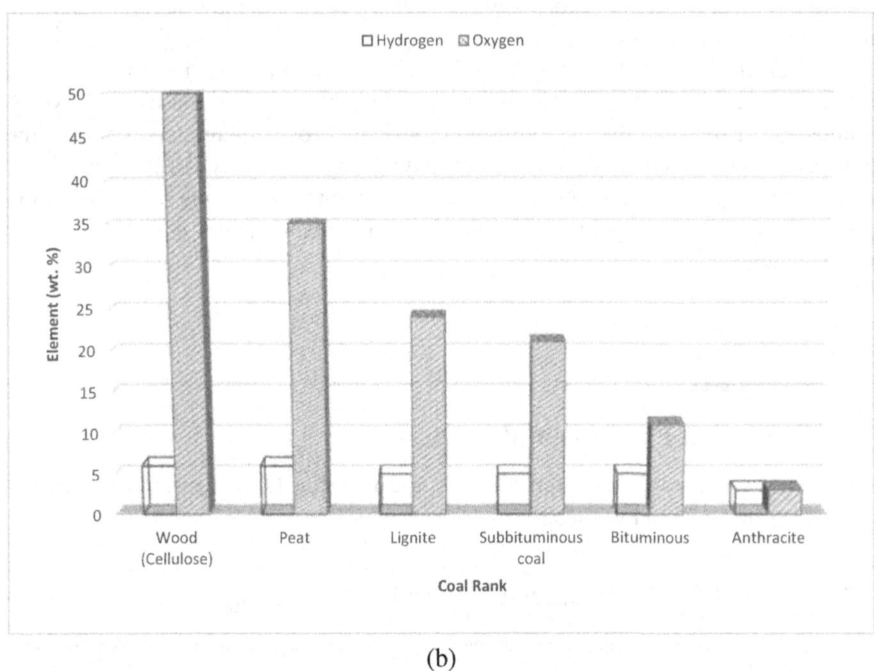

(b)

FIGURE 4.1
(a) Variation of carbon content with coal rank and (b) variation of hydrogen and oxygen contents with coal rank.

During coalification, the coal loses its oxygen while it is enriched in carbon as maturation progresses. It is worth noting that it appears that there are two forms of carbon – base carbon and carbon in volatile matter – present in coal which behave differently (Johnson, 1974; Wen and Heubler, 1965). Volatile carbon is a product of thermal pyrolysis, while base carbon is the remaining char left after devolatilization (Probstein and Hicks, 1982). In addition to carbon, hydrogen, and oxygen, coal also contains nitrogen and sulfur. These are undesirable impurities that need to be removed or reduced prior to the conversion process, especially when the process involves the use of catalysts.

Other important coal properties that are critical to coal conversion to hydrogen and other hydrogen-based products are the heating value and the proximate analysis (i.e., the concentration of moisture, volatile matter, fixed carbon, and ash) of the coal. The volatile matter, which is a product stream of coal devolatilization, is primarily made up of light oils, hydrocarbon gases, hydrogen, carbon monoxide, water, and tar. Low-rank coals such as lignite and subbituminous coal tend to contain high concentration of volatile matter and are consequently favored for coal-to-H_2 or coal-to-hydrogen-based products conversion process such as gasification and pyrolysis. It is worth noting that it appears that there are two forms of carbon – volatile matter carbon and base carbon – evolved when coal is heated (Johnson, 1974; Wen and Heubler, 1965). Volatile matter carbon is a product of thermal pyrolysis, while base carbon is the char residue left after devolatilization (Probstein and Hicks, 1982). Volatile matter carbon is said to be very reactive (Probstein and Hicks, 1982), and may therefore enhance reactivity of the conversion reactions.

Another important characteristic of coal of pertinence to its conversion is its polymeric nature. It is made of two building blocks – aromatics and hydro-aromatics. In-between the two building blocks are cross linkage, and flanking the two blocks are functional groups such as OH, –COO, –COOH, etc. This complex, unique structure changes when heat is applied to coal during coal conversion processes such as pyrolysis and gasification. This phenomenon is discussed later in this chapter.

4.3 Conversion Fundamentals

This section provides a brief description of some underline fundamentals of converting natural gas and coal to hydrogen.

4.3.1 Natural Gas-to-Hydrogen Conversion Processing

Regardless of the thermal/chemical technology used in producing hydrogen from natural gas, syngas is usually the precursor. This section describes the fundamentals of syngas production via the various technologies (SMR, ATR, POX, and natural gas (methane) pyrolysis).

4.3.1.1 Steam Methane Reforming Process

Steam methane reforming (SMR) is currently one of the most widely used technologies and at the same time least expensive processes for hydrogen production. Its advantage arises from the high efficiency of its operation and the low operational and production costs. The most frequently used raw materials are natural gas and lighter hydrocarbons, methanol, and other oxygenates. The general steam reforming reaction when any hydrocarbon is used as a feedstock is given by (Rostrum-Nielsen, 2001):

$$C_mH_n + mH_2O\ (g) \rightarrow mCO + (m + 0.5n)H_2 \qquad [4.1]$$

where m and n represent the number of atoms of carbon and hydrogen, respectively. The specific reaction when methane (contained in natural gas) is used as a feedstock for steam reforming is represented as follows (Kothari et al., 2004):

$$CH_4 + H_2O\ (+\text{Heat}) \rightarrow CO + 3H_2 \qquad [4.2]$$

Using Equation [4.1] and using appropriate values of m and n, Equation [4.2] can be written for other higher hydrocarbon gases (such as ethane, propane, butane, etc.) that are present in natural gas. The SMR step is an endothermic process, requiring addition of about 206 kJ/mole of external heat. Usually, the additional heat required is supplied by combusting a portion of the feedstock (natural gas in this case). Non-fossil-based external energy sources, such as nuclear, wind, solar, can be used to provide the required external heat. As well, Equation [4.2] is catalytic. A few catalysts have been used for SMR, including nickel-based catalyst (Kothari et al., 2004) and some precious metal-based materials, such as platinum and rhodium (Kalamaras and Efstathiou, 2013).

The SMR reaction (Equation [4.2]) is favored at low pressures, but is still done at high pressures (2.0 MPa, 20 atm). This is because high-pressure hydrogen is preferred most in the market. Also, the purification process step (pressure swing adsorption, PSA) that hydrogen is subjected to PSA before going to market work better at higher pressures. Water–gas shift (WGS) reaction is slightly exothermic (i.e., the reaction generates small heat). While the WGS reaction increases the hydrogen concentration of the gas product, it however, leads to production of carbon dioxide (CO_2) – a greenhouse gas, as shown in Equation [4.3]. A carbon capture and storage (CCS) process step can be integrated with the WGS system to capture any CO_2 produced. Since in practice, natural gas contains other hydrocarbon gases (ethane, propane, and butane) in addition to the predominant methane, a pre-reforming process is usually conducted. The pre-reforming is basically a fractionation process to convert those higher hydrocarbon gases to methane.

4.3.1.2 Partial Oxidation Process

As the name implies, partial oxidation is a process where the methane and other hydrocarbons contained in natural gas react with a limited amount of oxygen (less than the stoichiometric volume) in the air that is not enough to completely oxidize or combust the hydrocarbons to carbon dioxide and water. Because partial oxidation process is operated with less than the stoichiometric amount of oxygen, the reaction products contain primarily hydrogen and carbon monoxide (and nitrogen if the reaction is carried out with air rather than oxygen). Other components of the product mixture (typically in relatively small amounts) include carbon dioxide and other compounds, such as hydrogen sulfide (H_2S), carbonyl sulfide (COS), and unreacted hydrocarbon gases. The reactions for partial oxidation of methane with oxygen are, as given by Ogden et al. (1999) and Onozaki et al. (2006) are generally:

$$CH_4 + ½O_2 \rightarrow CO + 2H_2 \text{ (+Heat)} \quad [4.3]$$

$$CH4 + 2O_2 \rightarrow CO_2 + 2H_2O \quad [4.4]$$

Partial oxidation can also be carried out with steam (Rostrum-Nielsen, 2001), reaction of which is represented by Equation [4.5]:

$$CH_4 + H_2O\text{ (g)} \rightarrow CO + 3H_2 \quad [4.5]$$

Just like the case with the product gas mixture from SMR, in a WGS reaction, the CO in the partial oxidation product gas stream is reacted with water via the WGS reaction to yield more hydrogen. According to Semelsberger and co-worker (2004), the thermal efficiency of natural gas partial oxidation system can yield a thermal efficiency in the range of 60–75 percent. The non-catalytic process is conducted at a temperature range of 1300–1500°C and a pressure range of 3–8 MPa (Kalamaras and Efstathiou, 2013). Partial oxidation conducted in steam tends to produce a higher yield of hydrogen per unit quantity of feed methane, resulting in a higher H_2/CO ratio than when carried out in oxygen (H_2/CO ratio of 3/1 versus H_2/CO ratio of 1/1 or 1/2), as can be seen in equations [4.3–4.5].

Though partial oxidation of natural gas to produce syngas is usually a non-catalytic process, catalytic partial oxidation system (CPOX) can be employed to decrease the required operating temperature from 1300–1500°C range to about 700–1000°C range (Kalamaras and Efstathiou, 2013) if the coke formation propensity of the reactions (Song, 2003; Hohn and Schmidt, 2001; Krummenacher et al., 2003; Pino et al., 2002) can be prevented. According to Kalamaras and Efstathiou (2013), the catalysts typically used for natural gas CPOX are nickel- or rhodium-based. More details on the use of POX for converting methane and other hydrocarbon fuels to hydrogen can be found elsewhere (Krummenacher et al., 2003; Pino et al., 2002; Holladay et al., 2009; Semelsberger et al., 2004).

4.3.1.3 Autothermal Reforming Process

Autothermal reforming (ATR) is a process by which methane is thermally reacted with oxygen and carbon dioxide (CO_2) or steam to produce syngas, and subsequently produce hydrogen. Methane is partially oxidized during this exothermic reaction because of the oxidation. A syngas with H_2/CO ratio of 1/1 is produced when ATR utilizes carbon dioxide, while an ATR with steam yields a syngas product stream of 2.5/1 H_2:CO ratio. Equations [4.6] and [4.7] describe the reaction when the ATR is conducted using CO_2 and steam, respectively:

$$2CH_4 + O_2 + CO_2 \rightarrow 3H_2 + 3CO + H_2O \quad [4.6]$$

$$4CH_4 + O_2 + 2H_2O \rightarrow 10H_2 + 4CO \quad [4.7]$$

As reported by the Scientific Committee on Oceanic Research (1983), the temperature and pressure at which the syngas comes out of the single-stage reactor are between 950–1100°C and as high as 100 bar, respectively.

As depicted by equations [4.6] and [4.7], ATR is a combination of both steam reforming (endothermic) and partial oxidation (exothermic) reactions. ATR, therefore, has the advantages of not requiring external heat and being simpler and less expensive than SMR.

4.3.1.4 Natural Gas Pyrolysis Process

Pyrolysis, which came from Greek words "pyro" (meaning fire or heat) and "lysis" (meaning separating), is the thermal decomposition of hydrocarbons at elevated temperatures, often in an inert atmosphere. In general, devolatilization usually occurs first (after removal of water) during pyrolysis of carbonaceous organic materials, during which volatile components (including hydrogen) are released, leaving behind carbon black (when natural gas is the feedstock) and char (in the case of solid feedstock, such as coal or biomass), which is mostly carbon in both cases as the residue. Pyrolysis is also considered the first step in the processes of gasification or combustion (Zhou et al., 2013, 2015). According to Baden Aniline and Soda Factory research group (BASF, 2020), pyrolysis also has promise in the conversion of gas (primarily methane) into hydrogen and other hydrogen-rich products. It is also capable of converting biomass into syngas (from which hydrogen can be produced) and biochar, and waste plastics into value-added products.

Basically, pyrolysis generally involves application of heat to the material above its decomposition temperature in inert atmosphere to break the chemical bonds between the compounds to form atoms or smaller molecules. In the case of hydrogen production through natural gas pyrolysis process, the chemical bond between the carbon and hydrogen atoms in the methane (the predominant constituent of natural gas) and other hydrocarbon gases that may be present in the natural gas is broken to release the hydrogen. As reported by Zhou et al. (2013), the following processes generally occur at different temperature ranges when natural gas is subjected to pyrolysis:

- Below and up to slightly above 100°C, water is driven off the feed natural gas. This process stage is an energy-consuming step.
- Between 100 and 500°C, thermal decomposition of the gas molecule occurs, where methane and/or other higher hydrocarbon molecules break down by breaking the bond between the carbon and hydrogen atoms.

- At 200–300°C, if oxygen has not evolved, the left-over oxygen can react with the carbonaceous residue, resulting in a highly exothermic combustion reaction. Once carbon combustion starts, the temperature rises quickly, thereby sustaining the pyrolysis process.
- At about 350°C, devolatilization occurs, when volatile products (which may include water, hydrogen, CO, and/or CO_2, as well as many organic compounds) may be released at a temperature of up to 500°C, depending on the composition of the natural gas (Zhou et al., 2013, 2015, 2017).

Generally, this process also absorbs energy. The non-volatile residues typically become richer in carbon and form large, disordered molecules.

Industrially, methane pyrolysis is conducted by bubbling methane (CH_4) up through a reactor vessel containing dissolved nickel catalyst at about 1070°C to breakdown the methane molecule into hydrogen gas and solid carbon, as depicted by the following equation (Colias, 2020):

$$CH_4\,(g) \rightarrow C\,(s) + 2H_2\,(g) \quad \Delta H^\circ = 74\ \text{kJ/mol} \tag{4.8}$$

This process, which allows the direct removal of hydrogen from natural gas, is considered suitable for commercial bulk hydrogen production (Green et al., 2012). Further research continues in several laboratories, including at Karlsruhe Liquid-metal Laboratory (Upham, 2017) and the chemical engineering laboratory at University of California – Santa Barbara (Clarke, 2020).

4.3.1.5 Plasma Reforming

Plasma technology, which is based on a simple physical principle, is a phenomenon that changes the state of matter when energy is supplied to it. That is, solids can go into liquid state, and liquids go into gaseous state. With even more enough energy added to a gas, the gas is ionized and goes into the fourth state of matter – the energy-rich plasma state. Plasma technology, which was first discovered in the 1920s, has found its application in many industries, including fuel and energy conversion processes such as combustion, pyrolysis, gasification, and reforming (Patun et al., 2007). EnerSol Technologies, Inc., has developed a plasma pyrolysis system (PEPS), which has been demonstrated for converting petroleum coke (a carbonaceous material) to syngas (Patun et al., 2007), from which hydrogen can be produced. In the EnerSol petroleum coke-to-syngas system, a non-transferred DC (direct current) arc plasma torch is utilized to achieve a high-temperature environment with very reactive species (free radicals), which reacts with the petroleum coke and oxygen, thereby enhancing and accelerating the conversion reactions (Patun et al., 2007). Plasma energy can similarly be used to enhance and accelerate reforming reactions of other carbonaceous hydrocarbon materials, such as natural gas, coal, and biomass. According to Paulmier and Fulcheri (2005), plasma reforming technologies have been developed to facilitate POX, ATR, and steam reforming, with most of the reactors being POX and ATR (Paulmier and Fulcheri, 2005). There are essentially two main categories of plasma reforming, namely, thermal and non-thermal (Paulmier and Fulcheri, 2005).

Plasma reforming reactions tend to be same as those in conventional reforming represented by equations [4.2–4.7] depending on the medium in which the reforming is conducted. The major difference is the source of energy and functional groups/free radicals (Kalamaras and Efstathiou, 2013). In the case of plasma reforming, the source of energy and free radicals is plasma (Bromberg et al. 1997, 1999; Hammer et al., 2004; Paulmier and Fulcheri, 2005). When plasma reforming is conducted in the presence of steam OH and O radicals and electrons are formed, both reduction and oxidation reactions occur.

Very high temperatures in the range of 2000–10,000°C with a high degree of control using electricity can be achieved by plasma system (Bromberg et al. 1997, 1999; Hammer et al., 2004; Paulmier and Fulcheri, 2005; Patun et al., 2007). The heat generated is dependent on reaction chemistry, while optimal operating conditions can be maintained over a wide range of feed rates and gas compositions (Kalamaras and Efstathiou, 2012). One of the attributes of plasma system is its compactness, which is made possible by its high energy density, and by the short residence times, which occurs because of less reaction times, as outlined by Kalamaras and Efstathiou (2012). Conversion efficiency of almost 100 percent is said to have been reported for production of hydrogen-rich gas product stream from plasma reforming of a variety of hydrocarbon fuels (such as gasoline, diesel, oil, biomass, natural gas, and jet fuel) (Bromberg et al., 1997, 1999). Some of the other attributes of the plasma operating conditions that favor plasma reforming over conventional reforming include high temperatures, high degree of dissociation, and substantial degree of ionization, all of which can be used to rapidly increase the thermodynamic reactions in the absence of a catalyst or without the provision of additional external energy source needed for endothermic reforming processes (Kalamaras and Efstathiou, 2012).

4.3.2 Coal-to-Hydrogen Conversion Processing

The two methods commonly used to produce hydrogen from coal are gasification and pyrolysis. The fundamentals of these methods are briefly discussed here.

4.3.2.1 Coal Gasification

Coal gasification is a process by which coal is reacted with steam, air, and/or oxygen to produce syngas – a mixture consisting primarily of CO, H_2, CO_2, CH_4, and H_2O (water vapor). Other media, such as carbon dioxide and hydrogen, have also been used to gasify coal. Historically, coal gasification was utilized for producing coal gas (also known as town gas), which was then used for heating and streetlighting. Currently, large-scale coal gasification has found its application primarily in electric power generation or for production of hydrogen, and as feedstocks to produce synthetic liquid fuels and hydrogen-based products such as ammonia.

Although in a broad sense, coal gasification is a process whereby coal is thermally converted to a gas product stream, its application here is focused to the gas phase and gas–solid chemical reactions taking place at high temperatures in the gasification reactor (Probstein and Hicks, 1982).

When coal is fed into a gasifier during gasification, it is subjected to heat, thereby leading to some chemical changes/reactions. The first set of reactions is pyrolysis and devolatilization during which the volatile matter components are produced, thereby leaving behind solid carbon known as char. The volatile matter contains hydrogen, which can be recovered at this time. Upon further heating, the char (carbon) is gasified by reacting it with air or oxygen in the presence of steam, thereby resulting in partial oxidation of the carbon to produce syngas (a mixture made predominantly of CO and H_2, and small quantities of other compounds). Partial oxidation is achieved by ensuring that less than stoichiometric amount of air or oxygen is supplied to avoid complete combustion of the carbon. As described by Silk et al. (2007), the main reactions occurring during the char gasification can be classified into two main stages – oxidation and reduction:

Oxidation – This is when char reacts with oxygen in an oxygen-deficient environment (i.e., less than stoichiometric oxygen/fuel ratio) to form CO and CO_2. These reactions, when the carbon and hydrogen atoms in the char are partially oxidized, are as follows:

$$C + 0.5O_2 \rightarrow CO_2 \qquad \Delta H = -123 \text{ kJ/mol} \qquad [4.9]$$

$$C + O_2 \rightarrow CO_2 \qquad \Delta H = -406 \text{ kJ/mol} \qquad [4.10]$$

$$H_2 + 0.5O_2 \rightarrow H_2O_2 \qquad \Delta H = -248 \text{ kJ/mol} \qquad [4.11]$$

Reactions [4.9–4.11] are exothermic reactions, which typically occur at high temperatures (>1000°C), and are needed to drive the endothermic reactions represented by equations [4.12–4.14] in the reduction stage of the process (Silk et al., 2007). The carbon may also undergo a reversible reaction with CO_2 in an oxygen-deficient environment to produce CO, as represented by Equation [4.12]:

$$C + CO_2 \rightarrow 2CO \qquad \Delta H = 160 \text{ kJ/mol} \qquad [4.12]$$

Reduction – This is a reaction under a reducing atmosphere which may be provided with water. Such a reaction under gasification conditions is the endothermic formation of CO and H_2, as represented by Equation [4.13].

$$C + H_2O_2 \rightarrow CO + H_2 \quad \Delta H = 119 \text{ kJ/mol} \qquad [4.13]$$

The water, or rather the steam, when fed into the gasifier, can be a source of oxygen and hydrogen that may react with the carbon to form CO_2 directly under some conditions (Probbstein and Hicks, 1982), as given in Equation [4.14].

$$C + 2H_2O = CO_2 + 2H_2 \qquad [4.14]$$

Equation [4.14] is an endothermic reaction, and therefore, requires an external source of heat.

In addition to the aforementioned reactions, there may be some competing reactions such as those resulting into methane formations (equations [4.15–4.17]).

$$C + 2H_2 \rightarrow CH_4 \qquad \Delta H = -87 \text{ kJ/mol} \qquad [4.15]$$

$$CO + 3H_2 \rightarrow CH4 + H_2O \qquad \Delta H = -206 \text{ kJ/mol} \qquad [4.16]$$

$$3C + 2H_2O \rightarrow CH_4 + 2CO \qquad \Delta H = 182 \text{ kJ/mol} \qquad [4.17]$$

As mentioned earlier, the products most likely to be present in the product stream of gasification carried out in typical temperature range of 800–1500K are CO, CO_2, H_2, CH_4, and H_2O. The concentration of H_2 is usually enhanced through the WGS reaction that will be discussed later.

4.3.2.2 Coal Pyrolysis

As mentioned in Section 4.3.1.4, pyrolysis is the thermal decomposition of organic matter in the absence of oxidant (air or oxygen). During pyrolysis of carbonaceous materials such as coal, hydrogen-rich volatile matter is released, leaving behind carbon- and mineral matter-rich solid residue (char).

Pyrolysis is one of the processes of producing gaseous and liquid fuels (including hydrogen) from coal. It is usually the first stage in all coal conversion processes (gasification, liquefaction, and combustion) conducted at high temperature (Probstein and Hicks, 1982). As already mentioned, coal is made up of two building blocks (hydro-aromatics and aromatics), which are surrounded by functional groups and in-between which are cross-linkages. During pyrolysis, the functional groups are first released, followed by the breaking off the cross-linkages, and then followed by conversion of the hydro-aromatics to aromatics, as the application of heat to the coal progresses.

Although the study of pyrolysis has a long history dating back to the 18th century, a full and clear understanding of the fundamentals of the chemical and physical changes that occur during pyrolysis is still lacking (Probstein and Hicks, 1982). The main purpose of this section is therefore to attempt to shed some light that may be useful for the design and operation of syngas, and subsequently hydrogen production via pyrolysis. The chemical parameters of importance to evaluating the rate and quantity of volatile generated and product distribution at a given set of operating conditions include final pyrolysis temperature, heating rate, particle size distribution, and operating pressure (Anthony and Howard, 1976; Howard, 1981; Probstein and Hicks, 1982). Because coal is a polymeric material made up of a heterogeneous mixture of various organic and inorganic substances, the predictability of the yield and product distribution during pyrolysis is the least certain of all the carbonaceous materials. However, coal rank and properties play a critical role in determining volatile matter yield. For example, low rank coals (lignite and subbituminous) generally tend to have higher concentration of volatile matter and inherent inorganic exchangeable cation (such as sodium oxide, calcium oxide) that may influence pyrolysis product yield and distribution (Ogunsola and Lam,1993; Morgan and Scaroni, 1984; Morgan and Jenkins, 1984).

Thermal decomposition of coal begins at about 350°C, and increases with temperature up to 450°C, and decreases thereafter with further increase in temperature, as observed for lignite by Ogunsola and Azhakesan (1988. There are three pyrolysis rates – slow (100°C/s), medium pyrolysis (between 100 and 1000°C/s), and fast or flash pyrolysis (>1000–10,000°C/s). Slow pyrolysis tends to favor production of gaseous components, while fast pyrolysis favors production of liquids (Ogunsola and Azhakesan, 1988; Howard, 1981; Probstein and Hicks, 1982). Some details of the influence of coal constituents and pyrolysis operating conditions on volatile yield and composition are described elsewhere (Ogunsola and Azhakesan, 1987; Howard, 1981; Probstein and Hicks, 1982; Morgan and Jenkins, 1984; Morgan and Scaroni, 1984).

Yield of volatile matter generally decreases with increase in coal particle size (Howard, 1981). This is probably because of the heat transfer effect (i.e., faster heat transfer is expected in small coal particles than in large particles, Probstein and Hicks, 1982). As well, volatiles production during pyrolysis is expected to decrease with increase in pressure, but production of liquid products (tar and light oils) is favored at low pressure, while high pressure results in production of larger volumes of light hydrocarbon gases (Howard, 1981; Anthony and Howard, 1976; Probstein and Hicks, 1982).

4.3.3 Hydrogen Enrichment Process (Water–Gas Shift Reaction)

As mentioned earlier, product gas mixture generated from all the processes for converting either natural gas or coal to syngas (SMR, ATR, POX, pyrolysis, and gasification) usually contains significantly more carbon monoxide than hydrogen (Song, 2002; Kalamaras and Efstathiou, 2013). To increase the concentration of hydrogen in the syngas product stream, the carbon monoxide (CO) in the syngas can be converted to more hydrogen by reacting it with steam. This process reaction is the water–gas shift (WGS) reaction and is represented by Equation [4.5] (Muradov and Veziroglu, 2005; Silk et al., 2007):

$$CO + H_2O \rightarrow CO_2 + H_2 \quad \Delta H = -40\ kJ/mol \qquad [4.18]$$

The WGS reaction is slightly exothermic, giving off heat (about 40 kJ/mol). A high temperature is usually required for the WGS reaction to achieve fast reaction kinetics (Kalamaras and Efstathiou, 2013). Consequently, a decrease in hydrogen yield and high-equilibrium CO selectivity occurs, according to Kalamaras and Efstathiou (2013). The WGS reaction is also thermodynamically controlled, and it is favored at low temperatures (Silk et al., 2007; U.S. Department of Energy, 2005).

Although WGS rate of reaction is enhanced at high temperatures, it reduces hydrogen yield. As such, a two-stage WGS system (one at high temperature of about 350°C and the other at a low temperature of 190–210°C) is employed to achieve maximum conversion of CO to hydrogen (Silk et al., 2007). Both shift stages are catalytic, utilizing iron–chromium and copper–zinc catalysts, respectively, for the high-temperature shift and low-temperature shift.

Some additional reactions occurring within steam reforming processes have been studied (Xu and Froment, 1989; Hou and Hughes, 2001). Commonly the direct steam reforming (DSR) reaction is also included:

$$CH_4 + 2H_2O\,(g) \rightarrow CO_2 + 4H_2 \quad [4.19]$$

As these reactions by themselves are highly endothermic (apart from WGS reaction, which is mildly exothermic), a large amount of heat needs to be added to the reactor to keep a constant temperature. Optimal SMR reactor operating conditions lie within a temperature range of 800–900°C at medium pressures of 20–30 bar. High excess of steam is required, expressed by the (molar) steam-to-carbon (S/C) ratio. Typical S/C ratio values lie within the range 2.5:1–3:1 (Speight, 2020).

As reported by Rhodes et al. (2002), the high-temperature WGS conducted at about 310–450°C range in the presence of Fe_3O_4/Cr_2O_3 catalyst, led to CO reduction from 10 to 3 volume percent, while the low-temperature WGS system conducted at a temperature range of 180–250°C achieved a further reduction of CO to as low as 0.05 volume percent when catalyzed by a catalyst mixture of copper, zinc oxide (ZnO), and alumina (Al_2O_3).

The concentration of carbon monoxide in the product gas stream can be reduced further by subjecting the product gas to a CO selective methanation/oxidation (also known as preferential oxidation) process, details of which can be found elsewhere (Kalamaras and Efstanthiou, 2013; Song, 2002; Pietrogrande and Bezzeccheri, 1993). The preferential oxidation reaction is a catalytic process and the typical catalysts used are noble metals such as platinum, ruthenium, or rhodium supported on Al_2O_3 (Song, 2002; Pietrogrande and Bezzeccheri, 1993; Kalamaras and Efstathiou, 2013).

4.3.4 Product Upgrading/Purification

After the product gas mixture has been subjected to the WGS reaction to increase the hydrogen content of the product stream, the product gas mixture will be subjected to a purification/upgrading process step. This step involves application of pressure swing adsorption (PSA) and removal of final traces of impurities, such as CO_2, H_2S, and other impurities, leaving essentially pure hydrogen as the final product. A brief description of some of the methods for gas purification is given further. More details on gas purification can be found elsewhere (Probstein and Hicks, 1982; Kohl and Riesenfeld, 1974).

4.3.5 Removal of Acid Gases

The two acid gases commonly found in product gas stream of natural gas and coal conversion processes are H_2S and CO_2. Of the two acid gases, H_2S is the most difficult one to remove. The major technological pathways for removing acid gases include absorption, adsorption, and chemical conversion.

Adsorption technique is simply a process in which the gaseous phase substance is transferred into a liquid phase through the phase boundary, whereby the absorbed substance is either dissolved physically in the liquid or react chemically with the fluid. Unlike the absorption technique entails the transfer of an undesirable impurity from gaseous phase to a solid phase. In the case of adsorption technique, the process involves concentration of the impurities on the surface of a solid adsorbent. The chemical process of removing impurities is usually catalytic and involve chemical conversion of the impurities into another compound that can be subsequently removed easily thereafter.

4.4 Technologies

As mentioned, fossil fuel processing technologies can be used to convert hydrogen-containing materials, such as natural gas and coal into a hydrogen-rich gas stream. Fuel processing of methane (natural gas) is the most common commercial hydrogen production technology. Most fossil fuels contain a certain amount of sulfur, the removal of which is a significant task in the planning of hydrogen-based economy. As a result, the desulfurization process will also be discussed. In addition, the very promising plasma reforming technology recently developed will also be presented.

As discussed in Section 4.3, the technological routes that can be employed for producing hydrogen from natural gas and coal include: (a) SMR, (b) POX, (c) ATR, (d) pyrolysis, plasma reforming, and (e) gasification. These technologies produce a great deal of carbon monoxide (CO). Thus, in a subsequent step, one or more chemical reactors are used to largely convert CO into carbon dioxide (CO_2) via the water-gas shift (WGS) and preferential oxidation (PrOx) or methanation reactions, which are described later. In addition to those three basic technologies, plasma reforming technology has been used to generate hydrogen from fossil fuels.

In these four technological pathways, hydrocarbon-based materials (such as natural gas or coal) are utilized as the feedstock, and almost all of which are at commercial stage, operating at efficiency ranging from 60 to 85 percent (Kalamaras and Efstathiou, 2013). SMR is among the cheapest and most widely used technologies for hydrogen production (Ogden et al., 1999). According to the Department of Energy (DOE) Office of Hydrogen and Fuel Cell Technology, steam methane reforming is an advanced and mature technology that accounts for most hydrogen produced today in the United States.

The various steps involved in producing hydrogen from natural gas and coal include:

- Feedstock cleaning/beneficiation processing to remove undesirable impurities (such as sulfur, carbon dioxide, moisture, and other impurities) that may be present in the feedstock (natural gas or coal).

- Feedstock-to-hydrogen process. For this book, these processes include SMR, ATR, POX, pyrolysis, and coal gasification.
- Product stream enrichment process step to increase hydrogen concentration of the product stream. This step involves conversion of CO to more hydrogen. This process is also known as WGS reaction, as mentioned earlier.
- Product upgrading/purification, which includes application of PSA and removal of final traces of impurities (such as carbon dioxide) to improve product quality.

The aforementioned process steps apply to all the various methods (SMR, ATR, POX, pyrolysis, and gasification) of producing hydrogen from natural gas and coal.

4.4.1 Feedstock Pretreatment

The various cleaning/beneficiation processes used depend on the feedstock. For natural gas, the cleaning/beneficiation processes may include the following:

- Condensate, natural gas liquids (NGL), and water removal to remove any condensate and moisture in the gas.
- Fractionation, which is basically to remove the higher hydrocarbon gases (ethane, propane, butane).
- Removal of mercury can be done by using molecular sieve and activated carbon.
- Removal of acid gas such as H_2S and CO_2 – Some of the acid-gas treatment processes that can be applied may include amine treatment, Benfield process, sulfinol process, etc.
- Removal of sulfur – The most developed and commercialized desulfurization technologies are those that can catalytically convert organosulfur compounds. These include conventional hydrodesulfurization (HDS), hydrotreating with advanced catalysts and/or reactor designs, froth flotation, electrochemical processing, leaching methods, and a combination of hydrotreating with some additional chemical processes to maintain fuel specifications (Ogunsola, 1983; Ogunsola and Osseo-Asare, 1987; Babich and Moulijn, 2003; Song, 2003).
- Nitrogen removal, which can be done via cryogenic process, hydrodenitrogenation, absorption, and adsorption.

The first four beneficiation process steps are specific to natural gas. However, it is essential to de-water coal before entering the gasifier. Typically, well-head natural gas is passed through a natural gas processing plant to remove all the undesirable impurities and to bring it up to the required quality before sending it to the market for various uses, including the conversion processing plants such as SMR, POX, and ATR. The major impurity of concern in coal is sulfur. There are commercially available coal desulfurization technologies, details of which can be found elsewhere (Ogunsola, 1983; Ogunsola and Osseo-Asare, 1987; Meyers, 1977; Aplan and

Luckie, 1982; Wheelock, 1977; Song, 2003; Hernandez-Maldonado and Yang, 2002; Babich and Moulijn, 2003).

4.4.2 Conversion Technologies

The technological pathways described in this section for producing hydrogen from natural gas are SMR, POX, ATR, plasma reforming, coal gasification, and coal pyrolysis for generating hydrogen from coal.

4.4.2.1 Steam Methane Reforming Technology

Steam methane reforming (SMR) process basically involves reacting methane with high-temperature steam (700–1100°C (973–1380K)), at a pressure of up to 3.5 MPa (Kothari et al., 2004) to produce syngas – a mixture of hydrogen and carbon monoxide (CO). The whole SMR process involves a series of steps (including desulfurization, fractionation, steam reforming, hydrogen enrichment (WGS process), product upgrading/purification), as shown in Figure 4.2.

The network of reactions governing methane reforming has been described in Section 4.3.1.1. The whole SMR process is conducted in several stages as given before. In the first stage, the raw natural gas feedstock is subjected to a desulfurization process to remove any sulfur left in the natural gas. This is essential to avoid

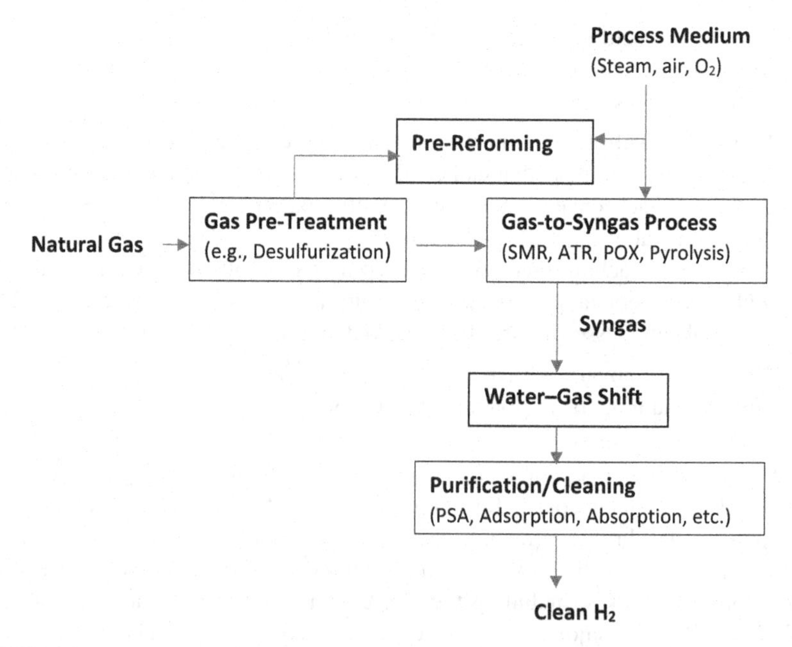

FIGURE 4.2
Simple flow diagram of overall gas-to-hydrogen process.

poisoning of the catalyst that is used in some of the downstream reforming reactions. The second stage is to remove the higher hydrocarbon gases (such as ethane, propane, butane, and pentane) that may be present in the natural gas feedstock. This stage is known as the pre-reforming or fractionation process. The next stage is the actual steam reforming stage, where the pretreated feedstock is fed, along with steam into a tubular catalytic reforming reactor (Song et al., 2007). The steam then reacts with the methane in the catalytic reformer over a nickel-on-alumina catalyst at 750–800°C (Silk et al., 2007) to produce syngas (a mixture of predominantly CO and H_2, along with lower concentration of CO_2).

As mentioned, the reforming reaction is an endothermic reaction, thereby requiring an external source of heat to sustain itself. This requirement is met by passing oxygen or air into the reforming reactor, whereby a combustion reaction between oxygen or air and additional source of the natural gas and tail gas from the SMR process occurs. The external source of heat required to sustain the endothermic reaction can also be obtained from other sources, including nuclear or other non-fossil fuels. As mentioned, the syngas product gas stream generated in the reforming stage is a mixture of CO and H_2. Because the residence time in the reforming reactor is extremely short, it is possible to obtain an extremely high yield of syngas (Silk et al. 2007). To obtain optimum hydrogen production, the CO needs to be converted to hydrogen to increase the hydrogen concentration of the product stream. This is achieved in the next stage of the process, which is called the hydrogen enrichment stage through a WGS reaction (Equation [4.18]). In this stage of the process, the cooled product syngas is fed into the WGS reactor where CO is catalytically converted to mostly hydrogen, and CO_2, by reacting steam with CO (Muradov and Verziroglu, 2005). As mentioned, the steam reforming reaction is a catalytic process. As such, the fed natural gas must be free of sulfur-containing compounds in to prevent catalyst deactivation.

Precious metal-based (platinum or rhodium) and non-precious metal-based catalysts (nickel) are typically used for SMR. Conventional SMR is rarely limited by kinetics (Rostrup-Nielsen, 2001), which enables the use of less expensive nickel catalysts for industrial application. According to Kalamaras and Efstathiou (2013), the propensity of CO_2 emission from SMR is dependent on the atomic H/C ratio in the feed, which is inversely proportional to the CO_2 emission potential. Efficiency of industrial SMR process ranges between 70 and 80 percent (Sorensen, 2011).

4.4.2.2 Partial Oxidation

Partial oxidation is a non-catalytic process conducted at a temperature range of 1300–1500°C and a pressure range of 3–8 MPa in the presence of oxygen (Ogden et al., 1999; Onozaki et al., 2006) and possibly steam (Rostrumm-Nielsen, 2001). The governing chemical reactions for POX have been described in Section 4.3.2.

Like in the SMR, the whole POX process is carried out in several stages, which include feedstock cleaning, fractionation, the actual oxidation reaction, hydrogen enrichment, and product purification/upgrading. It is worth noting that there is no need to desulfurize the feed natural gas during the feedstock beneficiation stage because the POX process is generally non-catalytic. In the actual partial oxidation reactor, methane is partially oxidized by oxygen, which is supplied to the reactor at sub-stoichiometric level to produce syngas. The product syngas stream formed via partial oxidation is made up of CO, CO_2, H_2O, H_2, CH_4, hydrogen sulfide (H_2S), and carbon oxysulfide (COS). Being an endothermic process, an external source of heat is required to sustain the reaction. As mentioned, this is achieved by burning a portion of the gas. POX operational cost is found to be relatively less expensive than that of SMR (Kalamaras and Efstathiou, 2013). However, the subsequent conversion during POX process renders the technology more expensive (Kalamaras and Efstathiou, 2013).

In some instances, POX can be operated in the presence of catalysts (such as nickel- or rhodium-based catalysts), which is a process known as catalytic partial oxidation (CPOX) system. CPOX, being a catalytic process, requires a feedstock desulfurization step prior to being fed into the CPOX reactor to prevent undesirable catalyst poisoning. The use of catalyst allows the partial oxidation to be carried out at lower operating temperature (700–1000°C) range. However, this situation may lead to coke formation and presence of hot spots, which may consequently make it difficult to control the operating temperature and yielding safety concerns (Kalamaras and Efstathiou, 2013; Song, 2002). As reported by Semelsberger et al. (2004), POX thermal efficiency is typically in the range of 60–75 percent when natural gas is used as the feedstock.

4.4.2.3 Autothermal Reforming

Like SMR and POX, the whole ATR process scheme for producing hydrogen involves similar several stages, as described in the previous sections. ATR is simply a combination of SMR (a predominantly endothermic process) and POX (a predominantly exothermic process) (Joensen, 1974 and Rostrup-Nielsen, 2001). ATR has the advantages of being relatively cheaper, less complex, and independent of additional external source of heat compared with SMR.

The operating condition of ATR is determined by the end use of interest. For example, a high hydrogen production rate with low CO content. Reforming in steam medium tends to favor maximum hydrogen production efficiency and high quality (i.e., low CO content). However, steam-based ATR is endothermic, requiring additional external energy source for sustainability.

Other significant positive attributes of ATR over SMR process are that its (ATR) capability can be carried out very rapidly, of producing a larger amount of hydrogen than POX alone, favorable gas composition for the Fischer-Tropsch synthesis, and shorter start-up and shut-down times (Joensen and Rostrup-Nielsen, 2002). Hence, application of ATR technology in the gas-to-liquid fuel industry is expected to grow. ATR is also relatively compact, has lower capital cost, and has the potential for economies of scale (Wilhelm et al., 2001).

Thermal efficiency of ATR when natural gas is used as feedstock is in the range of 60–75 percent, which is comparable with that of POX (Wilhelm et al., 2001).

4.4.2.4 Plasma Reforming

As mentioned earlier, plasma technology has found its application in many industrial sectors, including fuel and energy conversion processes such as combustion, pyrolysis, gasification, and reforming (Patun et al., 2007). Plasma reforming technologies have been reported to have been developed to enhance POX, ATR, and SMR (Paulmier and Fulcheri, 2005).

Plasma reforming can be carried out thermally or non-thermally (Paulmier and Fulcheri, 2005), reaching process temperatures above 2000°C, which can be controlled with a high degree of precision (Kalamaras and Efstathiou, 2013). Plasma reactors are compact by virtue of their short residence time and intensive high temper temperature, and are versatile (i.e., they can accommodate a variety of feedstock materials). Their conversion efficiencies have been reported to be almost 100 percent (Bromberg et al., 1997, 1999). This is made possible through the inherent operating conditions of high temperatures, high degree of dissociation, and enormous degree of ionization. These attributes negate the need for the use of catalysts. According to Kalamaras and Efstathiou (2013), several advantages are derivable from plasma reformers over some other reforming technologies, which include

- simple, light, and compact design because of its high energy density,
- high conversion efficiencies, low cost brought about using simple metallic or carbon electrodes and simple power supplies,
- short response time (fraction of a second),
- amenability to a broad range of feedstock materials, and
- versatile application – the technology could be used to produce hydrogen for a variety of stationary and mobile applications, such as distributed and low-pollution electricity generation for fuel cells, on-board generation of hydrogen for fuel-cell-powered vehicles (Kalamaras and Efstathiou, 2013; Bromberg et al., 1997).

Plasma reformers are, however, dependent on electricity and are not easy to operate at high pressure, which is required for high-pressure systems such as ammonia production (Kalamaras and Efstathiou, 2013).

4.4.2.5 Gasification

Gasification is a technological pathway for producing hydrogen from carbonaceous raw material such as coal. Gasification is usually carried out in a reactor known as gasifier, where coal is reacted with oxygen (or air) and steam at high temperature and pressure to first produce syngas, from which hydrogen is finally produced, as mentioned previously in the chapter. A series of chemical reactions that occur in coal gasifier have been described in Section 4.3.

Although gasification is suitable for most solid carbonaceous materials, the most promising solid carbonaceous feedstock option for gasification technologies in the

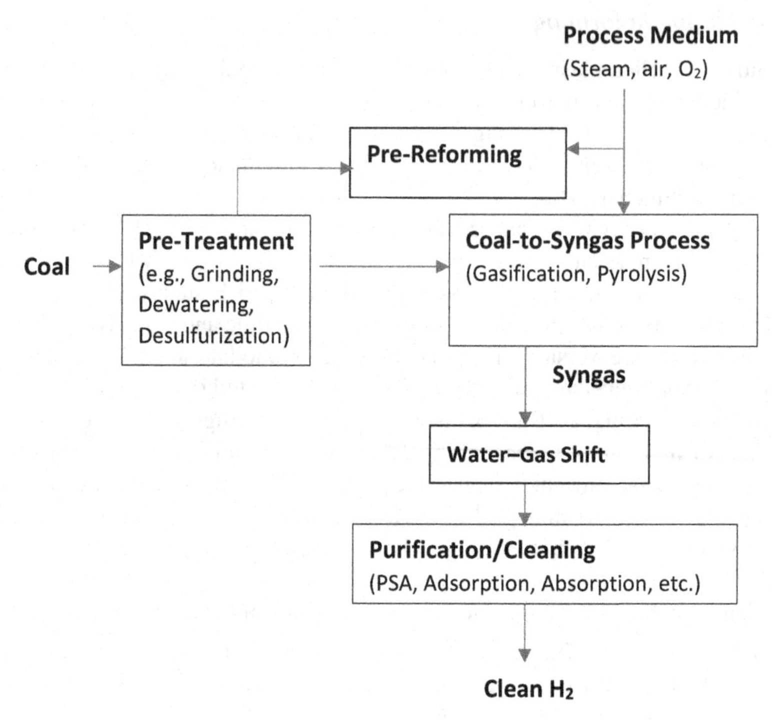

FIGURE 4.3
Simple flow diagram of a typical coal-to-hydrogen gasification process.

future is coal because of its (coal) global price stability and abundance. The future market situation/potential of hydrogen production through coal gasification technological pathway will be determined by many factors, including costs, reliability, availability, and maintainability, environmental considerations, efficiency, feedstock and product flexibility, energy security, public and government perception and policy, and infrastructure.

Figure 4.3 shows a simple flow diagram of a typical coal-to-hydrogen gasification process stream.

As shown in the flow diagram, the coal, after being subjected to appropriate cleaning/beneficiation process steps (drying, desulfurization, grinding, and sieving), is fed into the gasifier, in which it reacts with the appropriate gasifying medium (steam and oxygen) to produce syngas, through a series of chemical reactions, as discussed in Section 4.3. The oxygen is usually obtained from air which is passed through an air separation unit. There are four stages of events (drying, pyrolysis, gasification, and combustion) that typically occur in the gasifier reactor. The coal entering the gasifier first experiences a drying step by hot gases coming down the gasifier. The coal is next subjected to devolatilization step as heating progresses. At this stage, volatiles are evolved, leaving behind solid char, which is gasified in the gasification step brought about by the sufficiently high enough temperature brought about by the combustion step.

The syngas product gas stream produced in the gasifier is then sent to the WGS unit to increase the hydrogen content of the product stream at the expense of carbon monoxide content. Thereafter, the H_2-enriched gas product stream is sent to a gas stream purification/component separation stage to separate hydrogen gas from the stream and to remove undesirable impurities (such as sulfur and CO_2) from the final product (see Figure 4.3).

There are many types of commercial gasifiers with different designs and operational characteristics. However, they can be classified into three categories, based on the reactor-bed design. They are (a) fixed gasifiers (also known as moving-bed gasifiers), (b) entrained flow gasifiers, and (c) fluidized-bed gasifiers. While a detailed description of the major difference in the features and characteristics of these three categories of gasifiers can be found elsewhere (Phillips, 2006; U.S. Department of Energy, National Energy Technology Lab (NETL), 2002), a brief description of these three categories of gasification reactors, as reported in the U.S. DOE, NETL (2002) report, is presented in this section, as follows.

4.4.3 Fixed-Bed Gasifiers

Coal and oxidant are usually fed into the fixed- or moving-bed gasifiers counter-currently. That is, coal is fed through the top of the gasifier, while oxidants (steam and oxygen) are fed through the bottom of the reactor. The gasifier is commonly operated at moderate pressures of 25–30 atm. Coal in large particle size is slowly fed down the refractory-lined gasifier vessel where it moves slowly downward through the bed, while reacting with high oxygen content gas introduced at the bottom of the gasifier that is flowing counter-currently upward in the gasifier. Examples of commercial fixed-bed gasifiers include Lurgi dry ash gasifier and British Gas Lurgi gasifier.

As mentioned earlier, the top of the gasifier is the drying zone, where the entering coal is heated and dried while simultaneously cooling the product gas before exiting the gasifier. As the coal moves further down the reactor, it is further heated and devolatilization occurs as the higher-temperature gas descends through the carbonization regime, where tar and char are produced. The next zone is the gasification zone, where the char is gasified by reacting with steam and carbon dioxide (which are typical products of combustion). The final stage, which occurs near the bottom of the gasifier and at the inlet of the oxidant, is the combustion zone. It is in this zone that the heat required for the gasification is generated through the combustion of the remaining char.

There are two different modes of operating fixed-bed gasifiers – the dry-ash mode, an example of which is the Lurgi dry ash gasifier, and the slagging mode, an example of which is the British Gas Lurgi gasifier. Dry ash mode of operation requires that the gasifier is operated at temperatures below the ash-slagging temperature to prevent ash slagging in the reactor. This is achieved by operating at excess steam/char ratio. The ash, which is produced as a solid ash below the combustion zone, is cooled by the entering steam and oxygen or air. The slagging mode of operating a fixed-bed gasifier requires much less steam. Consequently, a much higher temperature is achieved in the combustion zone, thereby resulting in melting of the ash and production of slag. The discharge gas temperature, and consequently the slagging propensity of a fixed-bed gasifier is determined by the coal rank/moisture content. According to a report prepared by the Science Application International Corporation (SAIC) for the U.S. DOE NETL Gasification Technologies Program (2002), the discharge gas temperature of lignite, which has very high moisture content, is around 600°F (316°C), while that of a lower moisture-containing bituminous coal is >1000°F (538°C).

Lurgi dry ash gasification technology was first developed by Lurgi GmbH in the early 1930s to produce what was known as town gas, according to the U.S. DOE NETL (2002) report on gasification. As described in the report, the first commercial plant based on this technology was built in 1936 and further developed in the 1950s by Lurgi and Ruhrgas to handle bituminous coals in addition to the traditional lignite feedstock. Lurgi dry ash gasification technology has since been used worldwide to produce syngas, from which hydrogen can be obtained. This technology is the basis of such major projects as the Sasol synfuel plants in South Africa, and the Great Plains Synfuels Plant in North Dakota, and about 150 Lurgi gasifiers are said to be in operation, mainly in South Africa, China, and the United States (DOE/SAIC for the U.S. DOE NETL Gasification Technologies Program (2002)).

4.4.4 Entrained Flow Gasifier

Like in the fixed-bed gasifier, the oxidant used in the entrained flow gasifier is air or oxygen and/or steam. However, coal and oxidant are fed together into entrained flow gasifier, while counter-flow feeding is employed in the fixed-bed gasifier. This configuration in the entrained flow gasifier enables coal particles to be surrounded by the oxidant and steam as they (coal particles) flow through the gasifier in a dense cloud, experiencing turbulent flow down the gasifier. The turbulent and fast flow of particles gives rise to close contact between the coal and the oxidant, thereby resulting in rapid reactions and, consequently yielding high throughput and high carbon conversion efficiencies – 98–99.5 percent – (U.S. DOE NETL Gasification Technology report, 2002).

Entrained flow gasifiers are operated at high temperature and pressure with a very short residence time (in the order of few seconds). Like fixed-bed gasifiers, tar, oil, phenols, and other liquids are produced during the devitalization stage. These are decomposed into syngas, which is a mixture of H_2, CO, and small amounts of light hydrocarbon gases. By virtue of the rapid feeding of the small coal particles into entrained flow gasifiers, all types of coal (including caking coals) can be tolerated without any fear of agglomeration or caking. Both dry and slurries coal fines can be fed into the entrained flow gasifier. A lock hopper system is used when feeding dry coal particles, while high-pressure slurry pumps are used in delivering the feed in slurry form. Examples of entrained flow gasifiers include GE Energy (formerly known as Chevron), Texaco gasifier, Shell, Siemens, and others.

4.4.5 Fluidized-Bed Gasifier

A fluidized bed is a processing system that consists of a mixture of fluid and solids that behaves like fluid. The bed is basically considered to be a heterogeneous mixture of fluid and solids that can be represented by a single bulk density. This behavior is achieved by bubbling or passing fluid through a heated bed of fine particle sized solids at a velocity (fluidizing velocity) high enough to fluidize the solids and exhibit a fluid behavior without losing the solids (U.S. DOE NETL Gasification Technology report, 2002).

In fluidized beds, the solid particles are in very close contact with the fluidization medium (a gas or a liquid), thereby enabling good thermal transport inside the system and good heat transfer between the bed and its container/content. Consequently, a significant heat capacity and a uniform temperature across the bed are achieved. Fluidized beds have a wide industrial application, including gasification, and their basic properties that can be utilized for various applications include the following (U.S. DOE NETL Gasification Technology Report, 2002):

- Extremely high surface area contacts between fluid and solid per unit bed volume
- High relative velocities between the fluid and the dispersed solid phase
- High levels of intermixing of the particulate phase
- Frequent particle–particle and particle–wall collisions

All these are favorable attributes to enhance mass and heat transfer, high reaction rate, and high conversion efficiency in a fluidized-bed coal gasification.

According to Phillips (2006), a fluidized-bed coal gasifier is basically a back-mixed or well-stirred reactor with a consistent mixture of new coal particles, older unreacted coal particles, partially gasified coal particles, and fully gasified particles. Fed into the reactor is also a flow of gas (oxidant, steam, recycled syngas) that is fed through the bottom of the reactor (Phillips, 2006). The flow rate of the gas must be sufficient to keep the coal particles floating within the bed but not so high to entrain the coal particles out of the bed. However, as the particles are gasified, they will become smaller and lighter and will be entrained out of the reactor (Phillips, 2006). In a commercial fluidized-bed gasifier, a cyclone is typically installed downstream of the gasifier to capture the larger particles that are entrained out and then recycled back to the bed, when it happens (Phillips, 2006). Overall, the residence time of coal particles in a fluidized-bed gasifier is shorter than that of a moving-bed gasifier (Phillips, 2006).

During fluidized-bed coal gasification, coal particles of small size (<6 mm) enter at the side of the reactor, while steam and oxidant enter near the bottom with enough velocity to fully suspend or fluidize the reactor bed. Due to the thorough mixing within the gasifier, a constant uniform temperature is sustained in the reactor bed. The gasifiers normally operate at moderately high temperatures to achieve an acceptable carbon conversion rate in the range of 90–95 percent and to decompose most of the tar, oils, phenols, and other liquid by-products (U.S. DOE NETL 2002 Gasification Technologies Program report). To avoid clinker formation, fluidized-bed gasifier is usually operated at temperatures lower than ash fusion (Phillips, 2006). Relatively reactive coals and low rank coals are preferred for fluidized-bed gasification (Phillips, 2006). Examples of industrial fluidized-bed gasifiers include KBR Transport gasifier, High-temperature Winkler gasifier, U-Gas gasifier, and Great Point Energy gasifier.

4.4.5.1 Coal Pyrolysis

As mentioned, pyrolysis is one of the processes of converting coal to gaseous (including hydrogen) and liquid products (such as tar, oil), which can be reformed to produce hydrogen. Since it is usually the first stage in coal gasification process, the situation in the early stage of gasification, therefore, applies. The major product obtained during coal pyrolysis is char along with some liquid products (mostly tars), the yield of which depends on the configuration and design of the pyrolysis reactor and process variable. Like coal gasification, the three common types of reactors employed in pyrolysis technology are packed-bed reactors, entrained flow reactors, and fluidized-bed reactors. These reactor types have been described in Section 4.4.2.5. The variables that have influence on the yield, product distribution, and product quality include coal properties and operating conditions (pyrolysis temperature, pyrolysis pressure, pyrolysis rate, residence time). A discussion on the parametric effects on performance during coal pyrolysis has been given in Section 4.3.2.2.

Some of the known commercial pyrolysis technologies include the Char Oil Energy Development (COED) technology, Lurgi Ruhrgas technology, Occidental process technology, and the Rockwell technology (Probstein and Hicks, 1982). The coal rank of choice for all four processes is bituminous coal except for the Lurgi Ruhrgas process, which is designed to utilize subbituminous coal, and all the processes operate at a temperature around 500°C except for the Rockwell process, which operates at a relatively higher temperature (~900°C), according to Probstein and Hicks (1982). Similarly, the Rockwell process operates at a relatively higher pressure (about 3.5 MPa), while the other four processes are operated at a pressure of 0.33 MPa or less (Parsons, 1977; Probstein and Hicks, 1982).

During pyrolysis, the pretreated coal is fed into the pyrolysis reactor (pyrolizer) and heated to a predetermined pyrolysis temperature at a predetermined heating rate, which are based on the reactor type and design. The residence time, which is the time the coal is exposed to the pyrolysis reaction, is controlled by the volumetric feed rate. Typical products of pyrolysis are gases (including hydrogen), volatile matter, liquids (tar, oils, water), and char. The gaseous products, which are usually non-condensable, can be subjected to separation processes to recover hydrogen. The liquid products are condensed out and sent to a reformer to produce hydrogen. The hydrogen from both the gaseous and liquid pyrolysis products is upgraded and purified. Note, the char produced during pyrolysis can be gasified to produce syngas, from which hydrogen can be produced.

References

Anthony, D.B. and Howard, J.B. (1976). Coal Devolatilization and Hydrogenation. *AIChE J.*, 22, 625–656.

Aplan, F.F. and Luckie, P.T. (1982). Methods of Processing Coal to Remove Sulfur. *Earth Mineral Sci. Bull.* 61, 3.

Babich, L.V. and Moulijn, J.A. (2003). Science and Technology of Novel Processes for Deep Desulfurization of Oil Refinery Streams: A Review. *Fuel*, 82(6), 607–631.

Balat, M. and Balat, M. (2009). Political, Economic and Environmental Impacts of Biomass-Based Hydrogen. *Int. J. Hydrogen Energy*, 34(9), 3589–3603.

Balat, H. and Kirtay, E. (2010). Hydrogen from Biomass: Present Scenario and Future Prospects. *Int. J. Hydrogen Energy*, 35, 7416–7426.

BASF. (2020). BASF Researchers Working on Fundamentally New, Low-Carbon Production Processes, Methane Pyrolysis. United States Sustainability. BASF.

Bromberg, D., Cohn, R., and Rabinovich, A. (1997). Plasma Reformer-Fuel Cell System for Decentralized Power Applications. *Int. J. Hydrogen Energy*, 22(1), 83–94.

Bromberg, D., Cohn, R., Rabinovich, A., and Alexeev, N. (1999). Plasma Catalytic Reforming of Methane. *Int. J. Hydrogen Energy*, 24(12), 1131–1137.

Clarke, P. (2020). Dry Reforming of Methane Catalyzed by Molten Metal Alloys. *Nature Catalysis*, 3, 83–89.

Colias, M. (2020). Auto Makers Shift Their Hydrogen Focus to Big Rigs. *Wall Street Journal*. 26 October 2020.

DeLuchi, M.A. (1989). Hydrogen Vehicles: An Evaluation of Fuel Storage, Performance, Safety, Environmental Impacts and Cost. *Int. J. Hydrogen Energy*, 14, 81–130.

Green, R.A., Adams, R.W., Duckett, S.B., Mewis, R.E., Williamson, D.C., and Green, G.G. (2012). The Theory and Practice of Hyperpolarization in Magnetic Resonance Using Parahydrogen. *Prog. Nucl. Magn. Reson. Spectrosc.*, 67, 1–48.

Hammer, T., Kappes, T., and Baldauf, M. (2004). Plasma Catalytic Hybrid Processes: Gas Discharge Initiation and Plasma Activation of Catalytic Processes. *Catalysis Today*, 89(1-2), 5–14.

Hassman, K. and Kuhme, H.M. (1993). Primary Energy Sources for Hydrogen Production. *Int. J. Hydrogen Energy*, 18(8), 635–640.

Hernandez-Maldonado, A.J. and Yang, R.T. (2002). Desulfurization of Liquid Fuels by Adsorption via Complexation with Cu(I)-Y and Ag-Y Zeolites. *Indust Eng. Chem. Res.*, 42(1), 123–129.

Hohn, K.L. and Schmidt, L.D. (2001). Partial Oxidation of Methane to Syngas at High Space Velocities over Rh-Coated Spheres. *Appl. Catal.*, 211(1), 53–68.

Holladay, J.D., Hu, J., King, D.L., and Wang, Y. (2009). An Overview of Hydrogen Production Technologies. *Catalysis Today*, 139(4), 244–260.

Hou, K. and Hughes, R.R. (2001). The Kinetics of Methane Steam Reforming Over a Ni/α-Al_2O_3 catalyst. *Chem. Eng. J., Frontiers Chem. Reaction Eng.*, 82(1), 311–328.

Howard, J.B. (1981). Fundamentals of Coal Pyrolysis and Hydro-pyrolysis. In *Chemistry of Coal Utilization, Second Supplementary*, M.A. Elliott (Ed.), pp. 665–784. Wiley: New York, USA.

Johnson, J.L. (1974). Kinetics of Bituminous Coal Char Gasification with Gases Containing Steam and Hydrogen. In *Coal Gasification*, L.G. Massey (Ed.), pp. 145–178. Advances in Chemistry Series No. 131, American Chemical Society: Washington, DC, USA.

Kalamaras, C.M. and Efstathiou, A.M. (2013). Hydrogen Production Technologies: Current Status and Future Development. Energy Conference, Vol. 2013, Paper No. 690627, Hindawi Publishing Corp, London, England.

Kashtan, Y.S., Nicholson, M., Finnegan, C., Ouyang, Z., Lebel, E.D., Michanowicz, D.R., Shonkoff, S.B.C., and Jackson, R.B. (2023). Gas and Propane Combustion from Stoves Emits Benzene and Increases Indoor Air Pollution. *Environ. Sci. Technol.*, 57(26), 9653–9663.

Kidnay, A. J. and Parrish, W. R. (2006). *Fundamentals of Natural Gas Processing*, p. 9. CRC Press: Boca Raton, FL, USA.

Kohl, A. and Riesenfeld, F. (1974). *Gas Purification.* 2nd Ed. Gulf Publishing Company: Houston, TX, USA.

Konieczny, A., Mondal, K., Wiltowski, T., and Dydo, P. (2008). Catalyst Development for Thermocatalytic Decomposition of Methane to Hydrogen. *Int. J. Hydrogen Energy*, 33(1), 264–272.

Kothari, R., Buddhi, D., and Sawhney, R.L. (2004). Sources and Technology for Hydrogen Production: A Review. *Int. J. Global Energy Issues*, 21(1-2), 154–178.

Krummenacher, J.J., West, K.N., and Schmidt L.D. (2003). Catalytic Partial Oxidation of Higher Hydrocarbons at Millisecond Contact Times: Decane, Hexadecane, and Diesel Fuel. *J. Catal.*, 215(2), 332–343.

Mazloomi, K. and Gomes, C. (2012). Hydrogen as an Energy Carrier: Prospects and Challenges. *Renewable Sustain. Energy Rev.*, 16, 3024–3033.

Meyers, R.A. (1977). *Coal Desulfurization.* Marcel Dekker: New York City, NY, USA.

Morgan, B.A. and Jenkins, R.G. (1984). Role of Exchangeable Cations in Rapid Pyrolysis Lignite. In *Chemistry of Low Rank Coals*, Schobert, H.H. (Ed.). ACS Symposium Series 264, p. 214. American Chemical Society: Washington, DC.

Morgan, M.E., Jenkins, R.G., and Walker, P.L., Jr. (1981). Inorganic Constituents of American Lignites. *Fuel*, 60, 189.

Morgan, B.A. and Scaroni, A.W. (1984). Cationic Effects During Lignite Pyrolysis and Combustion. In *Chemistry of Low Rank Coals*, Schobert, H.H. (Ed.). ACS Symposium Series 264, p. 255. American Chemical Society: Washington, DC.

Muradov, N. Z. and Veziroglu, T. N. (2005). From Hydrocarbon to Hydrogen-carbon to Hydrogen Economy. *Int. J. Hydrogen Energy*, 30(3), 225–237.

Natural Gas.Org. (2014). Composition of Natural Gas: Understanding Its Key Element. www.naturalgas.org.

Ogden, M., Steinbugler, M.M., and Kreutz, T.G. (1999). Comparison of Hydrogen, Methanol and Gasoline as Fuels for Fuel Cell Vehicles: implications for Vehicle Design and Infrastructure Development. *J. Power Sources*, 79(2), 143–168.

Ogunsola, O.M. (1983). Physico-Chemical Mechanisms of Coal Desulfurization by Aqueous Chemical Leaching, M.S. Thesis, The Pennsylvania State University.

Ogunsola, O.I. and Azhakesan, M. (1988). Flash Pyrolysis of Nigerian Lignite in a Fluidized Bed Reactor. *Fuel*, 67, 1008–1011.

Ogunsola, O.I. and Lam, W.W. (1993). Mineral Composition of Nigerian Coals. *Fuel Sci. and Technology Int.*, 11(10), 1319–1329.

Ogunsola, O.M. and Osseo-Asare, K. (1987). The Electrochemical Behavior of Coal Pyrite 2: Effects of Coal Oxidation Products. *Fuel*, 66, 467–472.

Onozaki, K., Watanabe, T., Hashimoto, H.S., and Katayama, Y. (2006). Hydrogen Production by the Partial Oxidation and Steam Reforming of Tar from Hot Coke Oven Gas. *Fuel*, 85(2), 143–149.

Parsons, R.M. (1977). Coal Liquefaction Process Research Survey, R&D Interim Report No. 2 Data Source Book. Oak Ridge National Laboratory Report No. ORNL/Sub-7186/13, U.S. Department of Energy, Washington, DC, USA.

Patun, R., Ramamurthi, J., Vetter, M., Hartstein, A., and Ogunsola, O.I. (2007). Clean Fuels Production Using Plasma Energy Pyrolysis System. In *Ultraclean Transportation Fuels*, Ogunsola, O.I. and Gamwo, I.K. (Eds.), pp. 29–42. American Chemical Society Series 959, Oxford University Press: Oxford, England.

Paulmier, T. and Fulcheri, L. (2005). Use of Non-Thermal Plasma for Hydrocarbon Reforming. *Chem. Eng. J.*, 106(1), 59–71.

Phillips, J. (December 2006). Different Types of Gasifiers and Their Integration with Gas Turbines [PDF 1.2MB]. National Energy Technology Laboratory (NETL).

Pietrogrande, P. and Bezzeccheri, M. (1993). Fuel Processing. In *Fuel Cell Systems*, Blomen, L.J.M.J. and Mugerwa, M.N. (Eds.), pp. 121–156. Plenum Press: New York, NY, USA.

Pino, L., Recupero, V., Beninati, S., Shukla, A.K., Hegde, M.S., and Bera, P. (2002). Catalytic Partial-Oxidation of Methane on a Ceria-Supported Platinum Catalyst for Application in Fuel Cell Electric Vehicles. *Appl. Catalysis A*, 225(1-2), 63–75.

Probstein, R.F. and Hicks, R.E. (1982). *Synthetic Fuels*. Chemical Engineering Book Series. McGraw-Hill, Inc.: New York, NY.

Rhodes, C., Williams, B.P., King, F., and Hutchings, G.J. (2002). Promotion of Fe_3O_4/Cr_2O_3 High Temperature Water Gas Shift Catalyst. *Catalysis Commun.*, 3(8), 381–384.

Rostrup-Nielsen, R. (2001). Conversion of Hydrocarbons and Alcohols for Fuel Cells. *Phys. Chem. Chem. Phys.*, 3, 283–288.

Scientific Committee on Oceanic Research. (1983). Algorithms for Computation of Fundamental Properties of Seawater.

Semelsberger, A., Brown, L.F., Borup, R.L., and Inbody, M.A. (2004). Equilibrium Products from Autothermal Processes for Generating Hydrogen-Rich Fuel-Cell Feeds. *Int. J. Hydrogen Energy*, 29(10), 1047–1064.

Silk, M., Ackiewicz, M., Anderson, J., and Ogunsola, O. (2007). Overview of Fundamentals of Synthetic Ultraclean Transportation Fuels Production. In *Ultraclean Transportation Fuels*, Ogunsola, O.I. and Gamwo, I.K. (Eds.), pp. 3–17. American Chemical Society Series 959, American Chemical Society: Washington, DC..

Song, C. (2002) Fuel Processing for Low-Temperature and High Temperature Fuel Cells: Challenges, and Opportunities for Sustainable Development in the 21st Century. *Catalysis Today*, 77(1-2), 17–49.

Song, C. (2003). An Overview of New Approaches to Deep Desulfurization for Ultra-Clean Gasoline, Diesel Fuel and Jet Fuel. *Catalysis Today*, 86(1–4), 211–263.

Song, H., Zhang, L., Watson, R.B., Braden, D., and Ozkan, U.S. (2007). Investigation of Bio-ethanol Steam Reforming over Cobalt-Based Catalysts. *Catalysis Today*, 129(3-4), 346–354.

Sorensen, B. (2011). *Hydrogen and Fuel Cells*. Academic Press: Cambridge, MA.

Speight, J.G. (2020). *The Refinery of the Future* (2nd ed.). Cambridge, MA: Gulf Professional Publishing. ISBN 978-0-12-816995-7. OCLC 1179046717.

Upham, D.C. (2017). Catalytic Molten Metals for the Direct Conversion of Methane to Hydrogen and Separable Carbon. *Science. Amer. Assoc. Advanc. Sci.*, 358(6365), 917–921.

U.S. Department of Energy, Energy Information Administration. (2020). Natural Gas Explained.

U.S. Department of Energy, National Energy Technology Lab. (December 2002). Major Environmental Aspects of Gasification-Based Power Generation Technologies [PDF]. Final Report, Prepared by SAIC for NETL Gasification Technologies Program.

U.S. Department of Energy Office of Fossil Energy and Carbon Management. (2005). Hydrogen from Coal Program Research, Development, and Demonstration Plan for 2004 Through 2015.

Wen, C.Y. and Heubler, J. (1965). Kinetic Study of Coal-Char Hydrogasification. *Ind. Eng. Chem. Process Des. & Dev.*, 4, 142–154.

Wheelock, T.D. (Ed.) (1977). *Coal Desulfurization: Chemical and Physical Methods*. ACS Symposium Series 64.American Chemical Society, Washington, DC.

Wilhelm, D.J., Simbeck, D.R., Karp, A.D., and Dickenson, R.L. (2001). Syngas Production for Gas-to-Liquids Applications: Technologies, Issues, and Outlook. *Fuel Process. Technol.*, 71(1–3), 139–148.

Xu, J. and Froment, G.F. (1989). Methane Steam Reforming, Methanation and Water-Gas Shift: I. Intrinsic kinetics. *AIChE J.*, 35(1), 88–96.

Zhou, H., Long, Y., Meng, A., Li, Q., and Zhang, Y. (2013). The Pyrolysis Simulation of Five Biomass Species by Hemi-Cellulose, Cellulose and Lignin Based on Thermogravimetric Curves. *Thermochem. Acta.*, 566, 36–43.

Zhou, H., Long, Y-Q., Meng, A., Li, Q., and Zhang, Y. (2015). Thermogravimetric Characteristics of Typical Municipal Solid Waste Fractions During Co-Pyrolysis. *Waste Manag.*, 38, 194–200.

Zuttel, A., Borgschulte, A., and Schlapbach, L. (2017). *Hydrogen as a Future Energy Carrier*. Darmstadt: Vch Verlagsgsellschaft Mbh.

5

Hydrogen Production from Biomass

5.1 Overview

As alluded to by Probstein and Hicks (1982), biomass is directly or indirectly derived from plants and is renewable in less than a period of about 100 years. Since animal wastes are derivable from plants either directly or indirectly through the food chain, they can also be biomass. By virtue of its chemical characteristic nature, biomass is a carbonaceous material consisting mainly of carbon and hydrogen atoms, the sources of which are respectively carbon dioxide and water, both of which are products of combustion (Probstein and Hicks, 1982). Carbon dioxide and water are then absorbed by plants and converted to carbohydrates through a process known as photosynthesis, and eventually to biomass. This process can be represented by the following overall chemical reaction (Probstein and Hicks, 1982):

$$ {}_{n}CO_2 + {}_{m}H_2O \rightarrow C_n(H_2O)_m + {}_{n}CO_2 \quad \Delta H^\circ = +470 \text{ kJ/mol} \qquad [5.1] $$

Biomass is attracting increased interest, as clean, renewable energy alternatives are being sought. Compared with other renewable resources, biomass is very flexible and versatile in use. It can be used for heat and power and can be converted to a variety of products including hydrogen and other hydrogen-based products such as ammonia and fuels. According to the U.S. Department of Energy, Energy Information Administration (EIA) (April 2023) Monthly Review, production and consumption of biomass grew from about 4805 and 1211 trillion Btu, respectively, in 2020 to about 5158 and 1870 trillion Btu, respectively, in 2022.

Because of its wide range of feedstocks, biomass has a broad geographic distribution, in some cases offering a least cost and near-term alternative (Milbrandt, 2005). Biomass is an abundant renewable resource that can be produced domestically, and it can be converted to hydrogen and other hydrogen-containing products via various techniques. According to a study sponsored by U.S. Department of Energy, Energy Efficiency and Renewable Energy Office of the Biomass Program (2011), it is anticipated that with improvements in agricultural practices and plant breeding, up to 1 billion dry tons of biomass could be available for energy use annually.

Biomass-to-hydrogen appears to have an important inherent climate change attribute. The net carbon emissions from biomass gasification can be relatively lower than that of coal because biomass, during its growth, tends to remove carbon dioxide from the atmosphere.

DOI: 10.1201/9781003348283-5

This chapter discusses the fundamentals of biomass-to-hydrogen conversion and enumerates the various technological pathways being used to produce hydrogen from biomass. The technologies of interest for the purpose of this chapter can be divided into two categories – thermal conversion processes (gasification and pyrolysis) and biochemical conversion (fermentation and anaerobic digestion). The main product of gasification of interest here is syngas, from which hydrogen can be extracted, while pyrolysis produces both gases and liquids. Fermentation conversion results mainly in biomass liquids, which are subsequently reformed to manufacture hydrogen. The focus of this chapter is mainly on gasification, pyrolysis, and reforming, as potential technological pathways for producing hydrogen from biomass.

A brief discussion of the properties of biomass that are relevant to its conversion is also given in this chapter.

5.2 Relevant Properties of Biomass

Biomass, a renewable organic resource, is available from a wide range of sources, such as animal wastes, municipal solid wastes, crop residues, short rotation woody crops, agricultural wastes, sawdust, aquatic plants, short rotation herbaceous species (e.g., switch grass), wastepaper, corn, and many others (Asadullah et al., 2002; Demirbas, 2006). However, the description of biomass properties is given here in relation to its (biomass) potential for conversion to hydrogen and hydrogen-based materials such as synthetic fuels through biochemical and thermal process routes. While hemicellulose, cellulose, and lignin are the three main components of a biomass, the three main sources of biomass are generally agricultural (mainly wood and grain), municipal solid waste, and animal waste (manure). Grain (which typically includes corn stover, wheat, rice, and barley) is about 25 percent cellulose and about 10 percent lignin. These plants are known to be rich in starch that can be hydrolyzed to fermentable products (such as ethanol) from that can be reformed to produce hydrogen (Probstein and Hicks, 1982). Similarly, hemicellulose and some of the cellulose contained in wood can be hydrolyzed to fermentable sugars that can be subsequently converted to hydrogen (Probstein and Hicks, 1982; Office of Technology Assessment, 1980). Hydrogen production from wood is mainly through gasification and pyrolysis. Wood typically contains about 80 percent (on average) volatile organic matter, from which hydrogen and hydrogen-containing substances (such as syngas (mainly hydrogen, CO, CH_4, CO_2) and liquids (tar oil, methanol)) can be produced during biomass pyrolysis (Probstein and Hicks, 1982).

The key chemical properties of biomass are reflected in the proximate and ultimate analyses of the material. Proximate analysis is a measure of the concentration of volatile matter, moisture, ash, and fixed carbon contained in the material, while the ultimate analysis entails the measure of the concentration of the organic carbon, hydrogen, nitrogen, sulfur, and oxygen in the materials. Also, of importance for biomass conversion to useful products is the calorific value of each of the components of the biomass.

The results of analysis of such chemical properties of representative biomass materials, as given by the Solar Energy Research Institute (1980), Office of Technology Assessment (1980), and Probstein and Hicks (1982), reveal the following:

- The four main components of a typical biomass are wood, grain (agricultural crops), animal wastes, and municipal solid wastes (MSW), which are mainly organic and non-organic components.
- Wood has the highest carbon and organic volatile matter contents and the highest calorific value of all the four components of a typical biomass.
- Wood and agricultural crops (grains) have higher gross calorific value (about 20 MJ/kg) than the other two key components of biomass.
- Wood seems to have relatively higher moisture content than the rest of biomass constituents.

Wood with its relatively high concentrations of carbon, organic volatile matter, and high gross calorific value, thereby has the potential to contribute the most towards producing high-quality product, and in relatively larger quantity. Agricultural wastes, which are generally crop harvesting products, have limited potential for commercial hydrogen production. These materials are used onsite for either animal feeds or burned for meeting onsite energy demands. Municipal solid wastes (MSW) are made up of organic and non-organic components. The organic portion of MSW consists mainly of kitchen wastes, while the non-organic portion includes refuse, a high cellulose containing material, glass, and metals. Only the organic portion of MSW is suitable for biochemical conversion, while the refuse portion can be combusted to generate heat. Another drawback of MSW is its significant compositional variation with season, thereby limiting its use. Animal waste, which is usually derived from dairy and cattle rearing farms, has limited potential for large-scale commercial hydrogen manufacturing.

Partial oxidation of carbon during pyrolysis and gasification yields CO (a major component of syngas that is produced during reforming, pyrolysis, and gasification), which in turn is converted to hydrogen during the water–gas shift (WGS) reaction step. The combustion of carbon to form CO_2 during the conversion also generates the heat required to sustain the relevant process steps. The volatile matter generally contains gases (mainly CO, CO_2, CH_4, H_2), from which hydrogen can be separated, and liquids (tars, oils, which can be reformed to generate hydrogen. As such, it is reasonable to use biomass with high wood content for hydrogen production from the ultimate and proximate analyses viewpoint. Another advantage of wood is its relatively low concentration of ash compared to that of grain, MSW, and animal waste. However, wood appears to have relatively higher moisture content than the rest of biomass constituents, as observed earlier. In general, biomass contains more moisture than other potential feedstocks for hydrogen production. As noted in the Office of Technology Assessment report (1980), the hemicellulose and approximately 25 percent of the cellulose in wood are easily hydrolysable to fermentable sugars, and consequently render it amenable to biochemical conversion.

In summary, the composition and properties of biomass play a role in determining the appropriate conversion technologies. The relatively high moisture, oxygen, hydrogen, and volatile matter contained in biomass, and its low heating value are the main properties of biomass that are of critical importance in its thermal conversion. According to Probstein and Hicks (1982), the high concentration of hydrogen and oxygen is attributable to the high yield of volatile matter, gases, and liquids during biomass pyrolysis.

5.3 Biomass Conversion Fundamentals

As mentioned, biomass can be converted to syngas, from which hydrogen can be produced, via gasification and pyrolysis, and to liquid-derived products (such as cellulosic ethanol, bio-oils, tars, or other liquid biofuels), which can be converted to hydrogen and other hydrogen-based products through various process technologies (such as reforming, fermentation, and anaerobic digestion). Like the natural gas-to-hydrogen and coal-to-hydrogen conversion processes described in Chapter 4, the process steps involved in converting biomass are as follows:

- Feedstock cleaning, which may involve drying, desulfurization, grinding, etc.
- Syngas production by reacting biomass with steam or appropriate oxidant at high temperature
- Hydrogen enrichment of syngas product stream through WGS reaction to produce more hydrogen
- Separation and purification of the hydrogen product, which can be achieved with the use of adsorbers/absorbers or special membranes

The various techniques produce some sort of intermediate products, which are further processed to manufacture hydrogen. These intermediate products include syngas and methanol (from gasification), medium Btu gas (from pyrolysis process technology), ethanol (from fermentation of grains and sugar crops), and methanol (from aqueous reforming) (Probstein and Hicks, 1982).

It is worth noting that gasification and pyrolysis fundamentals are presented in Section 4.3.1 of Chapter 4. Even though the emphasis of Section 4.3.2 of Chapter 4 is on coal and natural gas pyrolysis and gasification fundamentals, much of the discussion in that section applies to biomass. Discussion in this section is, therefore, mostly on aspects of biomass conversion fundamentals that are different from those of coal and natural gas.

5.3.1 Biomass Gasification

Like coal gasification, biomass gasification is a mature and commercially available technology pathway that thermally reacts biomass with steam and oxygen to generate syngas, and subsequently hydrogen and other products.

During biomass gasification, the organic-based carbonaceous materials are partially oxidized by reacting the biomass materials with a controlled amount of oxygen and/or steam at high temperatures (>700°C) to form syngas (a mixture of mainly carbon monoxide and hydrogen, and some carbon dioxide and methane). This reaction can simply be represented by Equation [5.2]:

$$C_6H_{12}O_6 + O_2 + H_2O \rightarrow CO + CO_2 + H_2 + \text{other species} \qquad [5.2]$$

In the unbalanced Equation [5.2], glucose is used (for simplicity) as a surrogate for cellulose, which is a major component of a typical biomass. The actual biomass is highly variable and complex in composition. The other species in the product side of the equation includes methane. The gasification reaction can be further simplified by representing the biomass with carbon, as represented by Equation [5.3]:

$$C + H_2O \rightarrow CO + H_2 \qquad [5.3]$$

Please note that all the reactions and equations described for coal gasification fundamentals in Section 4.3.2.1 of Chapter 4 also apply to biomass gasification.

The CO in the product stream is then subjected to a WGS reaction (Equation 5.4) by reacting it with water to form CO_2 and more hydrogen. As mentioned, hydrogen as the final product is then separated from the product stream of the WGS step with the use of adsorbers/absorbers or special membranes.

$$CO + H_2O \rightarrow CO_2 + H_2 \qquad [5.4]$$

Once the hydrogen content is enriched through the WGS reaction step, the product can then be upgraded catalytically with the use of calcined dolomite and/or nickel catalyst (Asadullah et al., 2002).

5.3.2 Pyrolysis

Pyrolysis or co-pyrolysis, which is basically thermal decomposition of carbonaceous material such as biomass in the absence of an oxidant (or rather in an inert atmosphere), is another promising technique for producing hydrogen from biomass and/or coal. During biomass gasification, the pretreated biomass feedstock is fed into the gasification reactor where it is heated in an inert atmosphere to a temperature range of about 500–900°C and a pressure range of about 0.1–0.5 MPa (Ni et al., 2006; Muradov, 2003; Demirbas and Arin, 2004; Demirbas, 2005; Zhagfarov et al., 2005; Sakurovs, 2003; Kalamaras and Efstathiou, 2013). The advantages of pyrolysis technique for converting biomass to hydrogen include fuel flexibility, relative simplicity and compactness, clean carbon byproduct, and reduction in CO_x emissions (Ni et al., 2006; Muradov, 2003; Demirbas and Arin, 2004; Demirbas, 2005; Zhagfarov et al., 2005; Kalamaras and Efstathiou, 2013). According to Muradov (2003) and as reported by Kalamaras and Efstathiou (2013), Equation [5.5] represents the general reaction occurring in the biomass pyrolysis reactor:

$$C_nH_m + \text{Heat} \rightarrow nC + 0.5\, mH_2 \qquad [5.5]$$

It is worth noting that all the process fundamentals and chemical reactions described for coal pyrolysis in Section 4.3.2.2 of Chapter 4 also apply for biomass pyrolysis, and they will not be repeated here to avoid duplication.

According to Kalamaras and Efstathiou (2013), the temperature range in which pyrolysis is conducted can be divided into three different regimes. These are namely low-temperature regime (up to 500°C), medium temperature regime (500–800°C), and high-temperature regime (>800°C). As mentioned in Chapter 4, pyrolysis can also be conducted at three different rates: slow, medium, and fast (also known as flash pyrolysis; Ogunsola and Azhakesan, 1988; Morgan and Jenkins, 1984). Fast pyrolysis tends to favor gaseous products, while liquid products are favored at slow pyrolysis rate (Ogunsola and Azhakesan, 1988).

5.3.3 Aqueous Phase Reforming

Aquatic plants, which include kelp from the ocean and freshwater plants (e.g., water hyacinth, algae, duckweed), can be converted to hydrogen through aqueous phase reforming (APR). Aqueous phase reforming is a technology being developed to convert oxygenated hydrocarbons or biomass carbohydrates to hydrogen or hydrogen-based products (Davda et al., 2003a, 2003b; Huber and Dumesic, 2006). APR reactions are usually carried out at very lower temperatures of about 220–270°C compared with conventional steam alkane reforming, which occurs at 600°C (Kalamaras and Efstathiou, 2013). The low-temperature APR reactions minimize unwanted decomposition reactions that usually occur with high-temperature carbohydrates reforming (Kalamaras and Efstathiou, 2013; Kabyemela et al., 1999; Eggleston and Vercellotti, 2000). Another advantage of the low-temperature operation of APR reactor is that the WGS reaction to convert CO to more hydrogen occurs at the same temperature, thereby enabling the two reactions to occur in one stage (Kalamaras and Efstathiou, 2013). Efficiency of about 60 percent has been reported for APR-based reforming of butane and hexane by Rozmiarek (2008).

5.4 Technologies

As mentioned, hydrogen could be produced by other methods than from fossil fuels (Chapter 4) and electrolysis (Chapter 3). This section describes some biomass-based technologies (e.g., gasification, pyrolysis, reforming, fermentation, and anaerobic

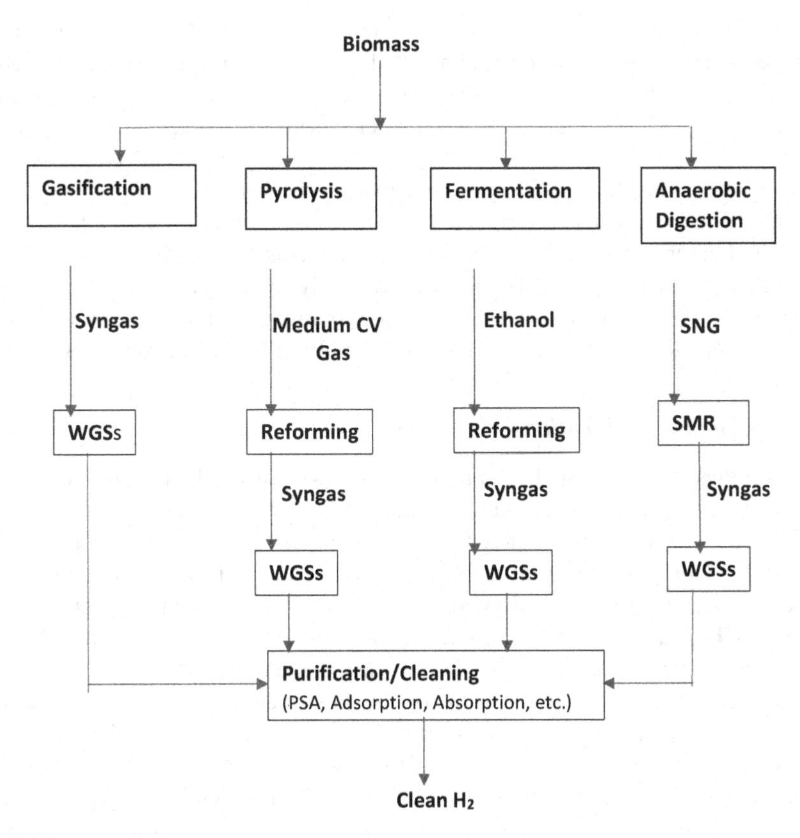

FIGURE 5.1
Simplified block diagram of biomass-to-hydrogen conversion processes.

digestion). The relevant biomass properties that make it amenable to hydrogen production and the relevant conversion fundamentals have been, respectively, described in sections 5.2 and 5.3 of this chapter.

By virtue of its properties and compositional characteristics, biomass conversion to hydrogen or hydrogen-based products can be achieved through two main categories of technologies. These are thermal, which include gasification and pyrolysis, and biochemical, which are mainly fermentation and anaerobic digestion. Figure 5.1 is a simplified block diagram of biomass conversion via these processes, the description of which is presented in this section.

5.4.1 Thermal Conversion Technologies

Pyrolysis product stream consists of gases, liquid, and solid, while gasification, which is carried out in the presence of oxidant, and at a higher temperature than pyrolysis, produces mainly gaseous product stream of better quality than that of pyrolysis. The superiority in the yield and quality of gasification over that of pyrolysis is probably a result of the oxidizing environment in which gasification is carried out compared with pyrolysis, which is conducted in an inert atmosphere. Another contributing factor is the absence of liquids and solids in the gasification product stream as they (liquid and solid char) have been reacted with the oxidant to form gaseous products.

It is worth noting that gasification and pyrolysis technologies are presented in Section 4.3.2 of Chapter 4. Even though the focus of Section 4.3.2 of Chapter 4 is on coal and natural gas pyrolysis and gasification technologies, much of the discussion in that section is applicable for biomass pyrolysis and gasification conversion technologies. Discussion in this section is therefore restricted to aspects of biomass conversion technologies that are different from those of coal and natural gas.

5.4.1.1 Biomass Gasification

During biomass gasification, the main goal is to convert as much of the biomass feedstock as possible to syngas product stream from which hydrogen can be obtained. As mentioned, the CO component of the syngas can be subjected to WGS process to increase the hydrogen content of the product gas stream, from which pure hydrogen can be separated through some downstream processes described in Chapter 4.

As noted, biomass contains very high moisture content, which tends to result in the production of gasification products with low heat content, which in turn will require a more energy intensive process step for its subsequent conversion to hydrogen. Hence, it is necessary to pre-dry biomass before being fed into a gasifier. However, biomass pretreatment step does not need to include desulfurization since biomass contains low sulfur content. Therefore, catalysts can be used for biomass gasification to improve its performance without a problem as the issue of catalyst poisoning and deactivation that is usually experienced with coal gasification will not be expected for biomass gasification and pyrolysis. Another couple of differences between biomass gasification and coal gasification, which are attributable to the difference between some properties of coal and biomass, are (Probstein and Hicks, 1982):

- The yield in gaseous products during biomass gasification is significantly higher in biomass gasification than during coal gasification because of the relatively high volatile matter content of biomass materials.
- Biomass-derived chars are significantly more reactive than those from coal, the effect of which when combined with the effect of higher volatile matter content connotes a lower oxygen and steam requirement for biomass gasification than for coal.

Like in the case with coal gasification, the biomass, after being subjected to appropriate cleaning/beneficiation process steps (drying, crushing, and sieving), is fed into the gasifier, in which it reacts with the appropriate gasifying medium (steam and oxygen) to produce syngas, through a series of chemical reactions, as discussed in Section 4.3. Biomass is subjected to the same four stages of events exhibited during coal gasification (drying, pyrolysis, gasification, and combustion), as described in Chapter 4.

As in the case with coal, the biomass fed into the gasifier is first exposed to a drying step by hot gases coming down the reactor. The biomass is thereafter devolatilized to produce volatile matters as heating progresses. At this stage, solid char is left behind, which is gasified in the gasification step resulting from the sufficiently high enough temperature resulting from the final step (the combustion step).

Just like the case with coal gasification, the syngas product gas stream produced in the gasifier is then sent to the WGS unit to convert the CO to more hydrogen, thereby increasing the hydrogen content of the product stream. Next, the H_2-enriched gas product stream is sent to a gas stream purification/component separation stage to separate hydrogen gas from the stream and to remove undesirable impurities (such as CO_2) from the final product.

Like coal, there are many types of commercial biomass gasifiers with different designs and operational characteristics. These gasifiers (as described in Chapter 4) can be classified into three categories, based on the reactor-bed design. They (again, as described in Chapter 4) are (a) fixed gasifiers (also known as moving-bed gasifiers), (b) entrained flow gasifiers, and (c) fluidized-bed gasifiers. The major differences in the features and characteristics of these three categories of gasifiers are summarized in the U.S. Department of Energy, National Energy Technology Lab (2002) report on Major Environmental Aspects of Gasification-Based Power Generation Technologies. The thermal efficiency of biomass gasification process is typically low, ranging from 35 to 50 percent (Kalamaras and Efstathiou, 2013) because of its high moisture content. Biomass gasification can be conducted in either a fixed-bed or fluidized-bed gasifier. However, its performance is typically better in a fluidized-bed reactor (Asadulla et al., 2002). A more detailed description of these categories of gasifiers can be found in Section 4.4.3 of Chapter 4 and elsewhere (Probstein and Hicks, 1982; U.S. DOE, NETL, Major Environmental Aspects of Gasification-Based Power Generation Technologies Final Report, 2002).

While gasification of biomass does not necessarily require the use of catalyst, it can, however, be conducted with the use of catalysts because they (catalysts) are not poisoned or deactivated due to its (biomass) low sulfur content. Alkali metal carbonate catalysts have been reported by Probstein and Hicks (1982) to enhance low-temperature wood gasification rate. Nickel and silica-alumina catalysts have also been shown to promote hydrogen-rich syngas production during steam wood gasification (Mitchell et al., 1980).

5.4.1.2 Pyrolysis

Biomass pyrolysis can basically be considered as gasification in the absence of oxidants. While biomass gasification is conducted at high temperatures (typically in the range of 800–1100°C or higher), pyrolysis is operated at a lower temperature (below about 600°C). Like in the case with coal pyrolysis, gases, liquids, and char are produced during biomass pyrolysis. However, smaller amounts of char are produced during biomass pyrolysis. The distribution of the products is determined by the pyrolysis temperature, heating rate, and biomass properties. Flash pyrolysis favors gas and liquids production, while char is the main product of slow pyrolysis. The gaseous products contain hydrogen that can be separated from the stream. Again, like in coal pyrolysis, the hydrocarbon liquid products can be reformed to produce hydrogen. It is worth noting that much of the discussions on coal pyrolysis fundamentals and technology can be extended to biomass. Discussion here is, therefore, concentrated on how biomass behaves differently during its pyrolysis.

The main biomass properties and their impacts on biomass pyrolysis have been described in Section 5.2. Worthy of mentioning is that the high generation of gaseous and liquids during biomass pyrolysis is a result of its (biomass) high oxygen and hydrogen contents. In summary, the yield and distribution of biomass pyrolysis products seem to be dictated by the biomass properties (oxygen, hydrogen, volatile matter, moisture) and the reactor operating conditions (heating rate, cooling rate and method, temperature, residence time). More details about the similarities and dissimilarities between coal pyrolysis and biomass pyrolysis can be found elsewhere (Probstein and Hicks, 1982; Solar Energy Research Institute, 1980).

5.4.2 Biochemical Conversion Technologies

For this book, biochemical conversion technologies, which are basically degradation or decomposition of selected constituents of biomass to various products, will be restricted to fermentation and anaerobic digestion. The main component of biomass targeted by fermentation is carbohydrates, which are decomposed to alcohols (mainly ethanol) in the presence of enzymes, while anaerobic digestion process technology is designed to decompose the organic material component of biomass to gaseous products (mainly CH_4 and CO_2) via anaerobic bacteria action. In producing hydrogen through biochemical conversion, the fermentation and anaerobic digestion processes will be integrated with a reforming process during which the respective products, ethanol from fermentation and methane from anaerobic digestion, will be reformed to produce syngas, from which hydrogen will be subsequently produced. The enzymes used in the fermentation process, which are basically catalysts for hydrolysis and hydrogenation reactions, are complex proteinaceous compounds produced by living cells (Probstein and Hicks, 1982).

5.4.2.1 Fermentation

Figure 5.2 depicts a simplified flow diagram of process steps involved in producing hydrogen from biomass via an integrated fermentation and reforming system. The process steps can be divided into two main classes. The first class consists of the steps involved in the production of ethanol through biomass fermentation. The second class of process steps include the steps that convert the ethanol to hydrogen via reforming.

The biomass is first subjected to a pretreatment stage, which includes grinding and drying. The pretreated feedstock is then sent into the fermentation reactor through

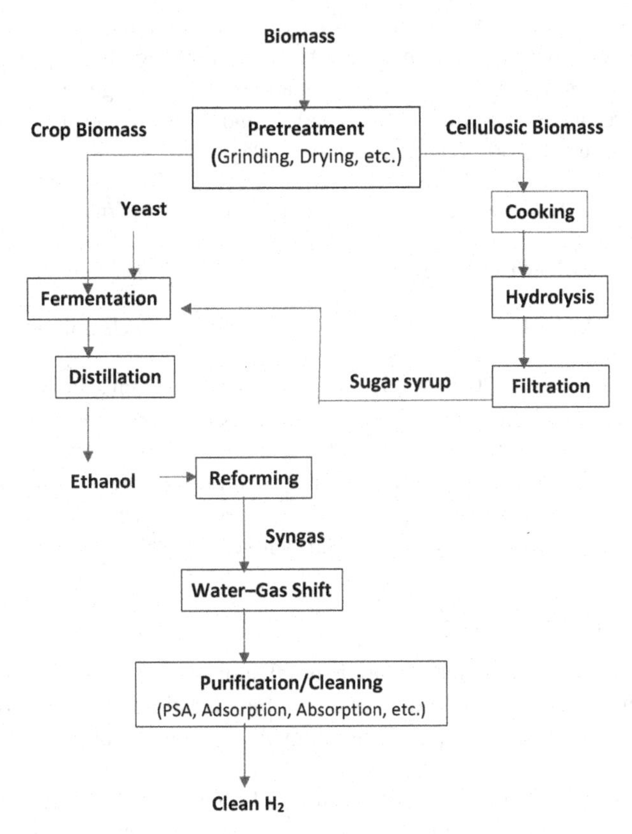

FIGURE 5.2
Simple block diagram of biomass-to-hydrogen via fermentation.

one of two routes, depending on the nature and type of the biomass material. The sugar crop-based part of the pretreated biomass is first sent into a sugar concentrating tank, where the sugar is concentrated. The starch or cellulosic part of the pretreated biomass is sent into a hydrolysis (saccharification) reactor, into which enzymes are added to hydrolyze the starch/cellulosic materials into sugar. Prior to entering the hydrolysis reactor, the starch crop/cellulosic material is fed into a tank of water, where the feed is suspended in boiling water. After being hydrolyzed to sugar, the sugar is then fed into the fermentation reactor along with the concentrated sugar obtained from the sugar crop part of the biomass. Yeast is added to the fermentation tank to aid the fermentation process. The product from the fermentation reactor is passed through a distillation unit to distil out the ethanol. More detailed description can be found elsewhere (Kelm, 1980; Probstein and Hicks, 1982). The ethanol can then be reformed to produce hydrogen in a way like steam methane reforming, as described in Chapter 4.

The process steps involved for producing hydrogen through ethanol reforming are same as those for SMR, and they include (a) steam reforming, (b) WGS reaction, and (c) selective carbon monoxide oxidation. Equations [5.6] through [5.8] represent the overall steam reforming reaction, WGS reaction, and selective carbon monoxide oxidation reaction, respectively (Haryanto et al., 2005).

$$C_2H_5OH + 3H_2O \rightarrow 2CO_2 + 6H_2 \quad \Delta H° = 174\ kJ/mol \qquad [5.6]$$

$$CO + H_2O \Leftrightarrow + CO_2 + H_2 \quad \Delta H° = 41\ kJ/mol \qquad [5.7]$$

$$CO + 0.5O_2 \rightarrow CO_2 \quad \Delta H° = 283\ kJ/mol \qquad [5.8]$$

In addition to H_2, CO_2, CO, and CH_4 are also generated from secondary reactions. The reforming process is catalytic. However, the major challenge is to develop highly active, selective, and durable catalysts for the reactions involved. The most active and selective catalysts commonly used for ethanol steam reforming are rhodium- and/or nickel-based. Palladium and platinum catalysts have also been reportedly used for ethanol reforming to produce hydrogen (Haryanto et al., 2005).

5.4.2.2 Anaerobic Digestion

While the main product of fermentation is ethanol (as discussed in the previous section), methane is the main product of anaerobic digestion. This process occurs in the following three steps (Office of Technology Assessment, 1980):

- Hydrolysis step – conversion of the insoluble constituents (carbohydrates, oils, and proteins) of the organic components of the biomass to soluble compounds (such as sugar and alcohols).
- Soluble compounds conversion step – this is the step where the soluble compounds are converted to intermediate products such as H_2, CO_2, esters, and fatty acids.
- Methanogenesis step – this is the final step where the intermediate products generated in the previous step are converted to methane. As an example, hydrogen methanation reaction can be represented by Equation [5.9] (Silk et al., 2007).

$$C + 2H_2 \rightarrow CH_4 \quad \Delta H° = -87\ kJ/mol \qquad [5.9]$$

A very simple schematic diagram of anaerobic digestion system is shown in Figure 5.3. As depicted in Figure 5.3 and like the fermentation process, the biomass feedstock is fed into a pretreatment step. The pretreated feedstock is then separated into two fractions – the starch/cellulosic (manure) fraction and the lignocellulosic (mainly crop residue) fraction. The lignocellulosic fraction is first sent into a hydrolysis reactor where acid or enzyme is added to hydrolyze the feed. The

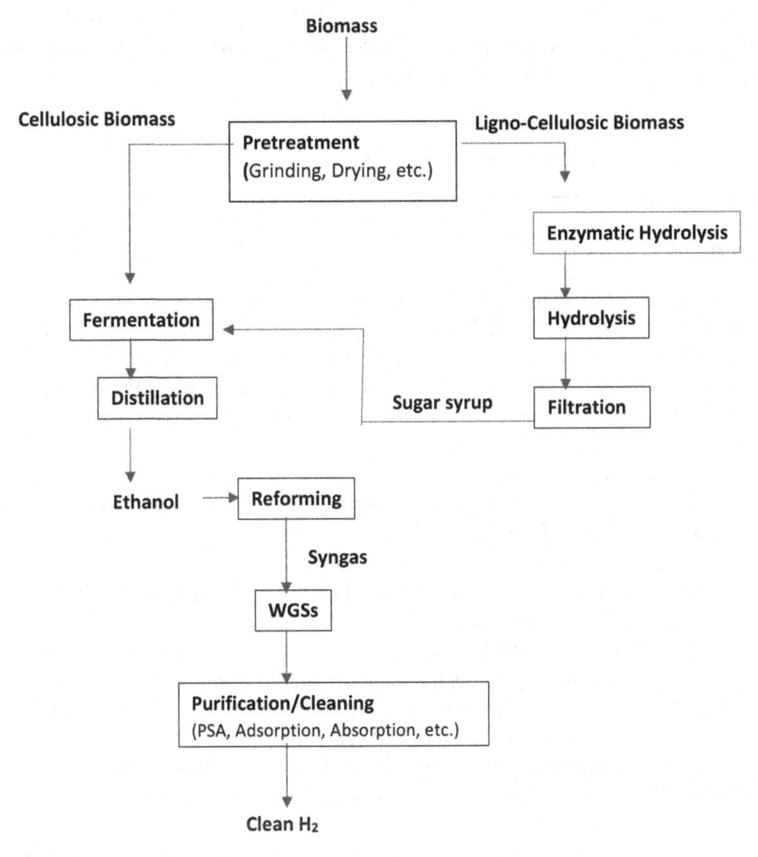

FIGURE 5.3
Simple block diagram of biomass-to-hydrogen via anaerobic digestion.

cellulosic fraction and the hydrolyzed lignocellulosic fraction are then fed into an equalization pond, to which water is added to homogenize the composition of the feed materials. The equalization residence time depends on the compositional variation of the feedstock fed into the equalization pond. The equilibrated material, along with nutrients, is fed into the digestion reactor, which provides the appropriate environmental conditions (oxygen-free, necessary bacterial growth, appropriate contact between bacteria and digestible organic feed materials, residence time, pH, and soak time) to convert the biomass efficiently and effectively into methane. The produced gaseous product stream is sent into an upgrading unit where impurities in the product stream are removed.

More details on anaerobic digestion of biomass can be found elsewhere (Probstein and Hicks, 1982). The produced methane is then fed into a steam methane reactor where it is reformed using the same process steps as described in Chapter 4 for natural gas reforming.

Anaerobic digestion system can be operated both in batch and continuous mode. It is generally a mild exothermic process, which may require some external heat source to sustain the process (Probstein and Hicks, 1982).

References

Asadullah, M., Ito, S.I., Kunimori, K., Yamada, M., and Tomishige, K. (2002). Energy Efficient Production of Hydrogen and Syngas from Biomass: Development of Low-Temperature Catalytic Process for Cellulose Gasification. *Environ. Sci. Technol.*, 36(20), 4476–4481.

Davda, R.R., Shabaker, J.W., Huber, G.W., Cortright, R.D., and Dumesic, J.A. (2003a). Aqueous-Phase Reforming of Ethylene Glycol on Silica-Supported Metal Catalysts. *Appl. Catalysis B*, 43(1), 13–26.

Davda, R.R., Shabaker, J.W., Huber, G.W., Cortright, R.D., and Dumesic, J.A. (2003b). A Review of Catalytic Issues and Process Conditions for Renewable Hydrogen and Alkanes by Aqueous Phase Reforming of Oxygenated Hydrocarbons Over Supported Metal Catalysts. *Appl. Catalysis B*, 56(1-2), 171–186.

Demirbas, A. (2005). Recovery of Chemicals and Gasoline-Range Fuels from Plastic Wastes via Pyrolysis. *Energy Sources*, 27(14), 1313–1319.

Demirbas, A. and Arin, G. (2004). Hydrogen from Biomass via Pyrolysis: Relationships Between Yield of Hydrogen and Temperature. *Energy Sources*, 26(11), 1061–1069.

Demirbas, M.F. (2006). Hydrogen from Various Biomass Species via Pyrolysis and Steam Gasification Processes. *Energy Sources A*, 28(3), 245–252.

Eggleston, G. and Vercellotti, J.R. (2000). Degradation of Sucrose, Glucose and Fructose in Concentrated Aqueous Solutions Under Constant pH Conditions at Elevated Temperature. *J. Carbohydrate Chem.*, 19(9), 1305–1318.

Haryanto, A., Fernando, S. Murali, N., and Adhikari, S. (2005). Producing Hydrogen Fuel for Fuel Cell Vehicles: Thermochemical Considerations. *Energy & Fuels*, 19, 2098.

Huber, G.W. and Dumesic, J.A. (2006). An Overview of Aqueous-Phase Catalytic Processes for Production of Hydrogen and Alkanes in a Biorefinery. *Catal. Today*, 111(1-2), 119–132.

Kabyemela, B.M., Adschiri, T., Malaluan, R. M., and Arai, K. (1999). Glucose and Fructose Decomposition in Subcritical and Supercritical Water: Detailed Reaction Pathway, Mechanisms, and Kinetics. *Industrial Eng. Chem. Res.*, 38(8), 2888–2895.

Kalamaras, C.M. and Efstathiou, A.M. (2013). Hydrogen Production Technologies: Current Status and Future Development. Energy Conference, Vol. 2013, Paper No. 690627, Hindawi Publishing Corp.

Kelm, C.R. (1980). Fuels and Chemical Feedstocks from Renewable Sources. *Int. Eng. Chem. Product Res. Dev.*, 19, 483–489.

Milbrandt, A. (2005). A Geographic Perspective on the Current Biomass Resource Availability in the United States. DOE National Renewable Energy Laboratory Technical Report, No. NREL/TP-560-39181.

Mitchell, D. H., Mudge, L.K., Robertus, R. J., Weber, S.L., Sealock, Jr., L.J. (1980). Methane/Methanol: By Catalytic Gasification of Biomass. *Chem. Eng. Prog.*, 76(9), 53–57.

Morgan, B.A. and Jenkins, R.G. (1984). Role of Exchangeable Cations in Rapid Pyrolysis Lignite. In *Chemistry of Low Rank Coals*, Schobert, H.H. (Ed.), p. 214. ACS Symposium Series 264.American Chemical Society, Washington, DC.

Muradov, N. (2003). Emission-Free Fuel Reformers for Mobile and Portable Fuel Cell Applications. *J. Power Sources*, 118(1-2), 320–324.

Ni, M., Leung, D.Y.C., Leung, M.K.H., and Sumathy, K. (2006). An Overview of Hydrogen Production from Biomass. *Fuel Process. Technol.*, 87(5), 461–472.

Office of Technology Assessment. (1980). *Energy from Biological Processes*. Vols. I, II, and III. A-C. U.S. Government Printing Office, Washington, DC, USA.

Ogunsola, O.I. and Azhakesan, M. (1988). Flash Pyrolysis of Nigerian Lignite in a Fluidized Bed Reactor. *Fuel*, 67, 1008–1011.

Probstein, R.F. and Hicks, R.E. (1982) *Synthetic Fuels*. Chemical Engineering Book Series, McGraw-Hill, Inc., New York.

Rozmiarek, B. (2008). Hydrogen Generation from Biomass-Derived Carbohydrates via Aqueous Phase Reforming Process. Presented at the Annual Merit Review Proceedings, J. Milliken (Ed.), pp. 1–6. Department of Energy, Washington, DC, USA.

Sakurovs, R. (2003). Interactions Between Coking Coals and Plastics During Co-Pyrolysis. *Fuel*, 82(15–17), 1911–1916.

Silk, M., Ackiewicz, M., Anderson, J., and Ogunsola, O. (2007). Overview of Fundamentals of Synthetic Ultraclean Transportation Fuels Production. In *Ultraclean Transportation Fuels*, Ogunsola, O.I. and Gamwo, I.K. (Eds.), pp. 3–17. American Chemical Society Series 959, Oxford University Press, Washington, DC.

Solar Energy Research Institute. (1980). A Study of Biomass Gasification. Vols. I–III, Report No. SEIR/7R-33-239, Golden, CO, USA.

U.S. Department of Energy, Energy Efficiency and Renewable Energy Office of the Biomass Program. (2011). A Report on U.S. Billion-Ton Update: Biomass Supply for a Bioenergy and Bioproducts Industry.

U.S. Department of Energy, Energy Information Administration. (April 2023). Renewable Energy Production and Consumption by Source. Monthly Review.

U.S. Department of Energy, National Energy Technology Lab. (December 2002). Major Environmental Aspects of Gasification-Based Power Generation Technologies [PDF]. Final Report, Prepared by SAIC for NETL Gasification Technologies Program.

Zhagfarov, F.G., Grigor'Eva, N.A., and Lapidus, A.L. (2005). New Catalysts of Hydrocarbon Pyrolysis. *Chem. Technol. Fuels Oils*, 41(2), 141–145.

6

Hydrogen Delivery and Transportation

6.1 Introduction

About 85 percent of hydrogen currently produced is utilized on-site, while the remaining 15 percent is transported in pipelines or trucks (International Energy Agency, 2019). The split between the two may change in the future as new opportunities become available. Typically, after production, the next step in the hydrogen value chain is to deliver the produced hydrogen either directly to the market through appropriate distribution network or to a storage facility, from where it will be distributed to various places of application and use. This chapter describes the various technologies that are available and are being developed for delivering and transporting hydrogen.

Hydrogen delivery infrastructure may consist of many technology pathways that are adequate for transporting hydrogen in various forms (gas, liquid, chemical). Gaseous hydrogen can be transported through pipelines and high-pressure tube trailers, while liquid hydrogen can be transported in tanker trucks. Chemicals such as ammonia may also be used as hydrogen carriers. In addition, there may be a need for a dispensing system along the hydrogen value chain. The nature in which the hydrogen is transported, stored, and used dictates the choice of dispensing technology used.

According to the U.S. Department of Energy Hydrogen Program Plan (2020), there are four major methods of delivering hydrogen at scale – gaseous tube trailers, liquid tankers, pipelines (for gaseous hydrogen), and chemical hydrogen carriers. These technological pathways are described in this chapter.

Tube trailers, which normally operate at high pressure, are used for transporting small amounts of hydrogen (typically less than 1 metric ton/day), as reported in the DOE Hydrogen Program Plan. Transportation of hydrogen via pipelines is used in areas with reliable, steady, and long-term demand of very large quantities of hydrogen. There are currently more than 2575 km (1600 miles) of dedicated pipelines transporting hydrogen in the United States (DOE Hydrogen Program Plan, 2020). Pipelines are the most energy efficient way of transporting hydrogen. This attribute is, however, negated by its high capital cost. Another positive attribute for pipeline option is the possibility of using existing natural gas pipelines to transport hydrogen. In this case, hydrogen can be co-mingled with natural gas and transported together in the same pipeline.

DOI: 10.1201/9781003348283-6

Hydrogen can also be transported in liquid form using liquid tankers especially to areas in regions where hydrogen demand is too small to necessitate pipelines. International hydrogen markets can also receive liquid hydrogen by loading the liquid tankers onto a ship. It is relatively safer to transport liquid hydrogen than gaseous hydrogen. However, liquid hydrogen requires a liquefaction facility where the gaseous hydrogen is chilled into a liquid at –253°C (–425°F) using liquid nitrogen. Although hydrogen liquefaction technology is a mature technology, it is, an energy-intensive process, thereby adding to the overall cost of transporting hydrogen in liquid form. Hence, the process is both capital- and energy-intensive.

Large quantities of hydrogen can also be transported by using chemicals to carry hydrogen. This novel technology is known as chemical hydrogen carriers. It is claimed to have potential to carry more hydrogen than tube trailers and at lower cost (DOE Hydrogen Program Plan, 2020). Chemical hydrogen carriers are materials (liquid or solid) that can chemically bond with hydrogen at low pressure and approximately ambient temperatures but can then release the hydrogen on demand (DOE Hydrogen Program Plan, 2020). They may be suitable for an unstable large market that does not necessitate a pipeline. Chemically carried hydrogen is cheaper to deliver than transporting gaseous or liquefied hydrogen. Commercial chemical hydrogen carrier technology development is still in its early stage. Examples of chemical hydrogen carriers include ammonia – a one-way carrier that does not release a by-product for reuse or disposal after the hydrogen is released, and methylcyclohexane/toluene whose by-products are typically returned for processing for reuse or disposal after the hydrogen is released.

Another important part of hydrogen delivery system is the dispensing and fueling system. Upon delivery either to the storage facility or to the place of use, hydrogen may have to be subjected to various additional processes such as compression, cooling, and/or purification, and dispensing/fueling, depending on the delivery and utilization methods. For example, hydrogen fueling stations with high-pressure compressors, storage vessels, and dispensers are required for hydrogen use.

6.2 Hydrogen Transportation Technologies

As indicated by Ball and Weeda (2016) and as mentioned earlier, there are several methods to transport hydrogen from its point of production to the end users, depending on the quantity to be transported. By virtue of its low energy density, hydrogen transportation over long distances may be costly. However, hydrogen can be transformed in various forms (such as compression, liquefaction, or incorporating it into bigger molecular compounds) to reduce the transportation cost. The methods by which hydrogen is transported can be classified into the following three main categories – road (for transporting gaseous hydrogen), pipeline (for gaseous and liquefied hydrogen), and ocean/shipping for transporting all three phases of hydrogen (gaseous, liquefied, and solid).

6.2.1 Transporting Hydrogen by Road

As mentioned, hydrogen can be transported by road either in compressed gaseous form or in liquid form. It can then be subsequently loaded onto a road truck or a trailer in appropriate containers – pressure vessels for compressed gaseous hydrogen and cryogenic liquid tanks for liquid hydrogen.

6.2.1.1 Transportation of Compressed Gaseous Hydrogen by Road

Road trucks can be used to haul pressure vessels filled with compressed gaseous hydrogen. There are four different types (types I, II, III, IV) of pressure vessels for handling hydrogen. While detailed description of the characteristics and uses of the various types of pressure vessels can be found in a review paper by Yang and co-workers (2023), a summary of the relevant characteristics and area of application of the four types of pressure vessels is given here as follows:

- **Type I pressure vessel** – Are fully metallic pressure vessels that are designed to operate at a pressure range of 15–30 MPa and are mainly for industrial gas storage.
- **Type II pressure vessels** – These are metallic pressure vessels hoop-wrapped with glass fiber composite that can operate under a wider range of pressure (10–90 MPa) than the Type I vessels. They are suitable for stationary high-pressure storage applications such as re-fueling stations.
- **Type III pressure vessels** – Are vessels made with full composite material wrapped with metal liner that are designed to withstand pressures higher than 100 MPa. They are mostly used for transport applications such as tube trailers, trailers, and on-board storage.
- **Type IV pressure vessels** – These are vessels made of fully composite inner materials (e.g., high-density polyethylene) with glass or carbon fiber designed to operate at a pressure range of 30–70 MPa. They are suitable for stationary low-pressure storage, tube trailers, trailers, and on-board storage applications.

It appears that types III and IV pressure vessels are used for transporting hydrogen, while the other two are only for storing hydrogen.

Once the pressure vessels are loaded with hydrogen, they are then loaded onto the truck. The number of pressure vessels loaded on the truck depends on the volume of hydrogen required at a particular destination and the length of the truck container. According to Yang et al. (2023), one truck can haul 200–1000 kg of hydrogen. Hexagon Purus GmbH (2022) indicated that a gas container, with lengths varying from 3 to 12 m consisting of 22–103 vertically installed Type IV pressure vessels can carry 240–1115 kg of hydrogen at a pressure of 50 MPa.

Another way in which hydrogen can be transported by road is by using the gas tube trailers that are used for transporting natural gas. For this approach, Type III vessels horizontally bundled together are usually used (Yang et al., 2023). The capacity of a tube trailer of Type III vessel is estimated at about 380 kg of hydrogen regulated at 25 MPa, compared with a capacity of 560–900 kg of hydrogen when the trailer is loaded with Type IV vessels, according to the U.S. Department of Energy, Office of Hydrogen and Fuel Cell Technology (2022). This option is limited to customers with small hydrogen demand of compressed gaseous hydrogen (CGH_2) located within a distance not more than 200 km (Galassi et al., 2012; Yang et al., 2023). According to Yang et al. (2023), the appropriate distances for transporting compressed gaseous hydrogen by road truck, liquid hydrogen by road truck, pipeline, and liquid hydrogen via shipping are up to 100, >500, up to 1000, and >1000 km, respectively.

6.2.1.2 Transportation of Liquid Hydrogen by Road

Another form in which hydrogen can be transported by road is liquid. The gaseous hydrogen is first sent to a liquefaction plant where it is subjected to a liquefaction process, which involves cooling of the gas to about –260°F at a pressure of about 30–70 MPa. Under these conditions, the hydrogen is turned into a liquid state. The liquid hydrogen can then be loaded into cryogenic tanks and be transported in trailers by road to the end users. As reported by Yang et al. (2023), the storage density of hydrogen is significantly increased when it is converted to a liquid state, making it possible for a much larger volume of liquid hydrogen to be transported per trailer than compressed gaseous hydrogen. The delivery of liquid hydrogen by trailers is reported in a case study by Reuß et al. (2021) to be particularly suitable for medium distances (>130 km). While it is safer to transport liquid hydrogen, it is, however, more energy intensive and expensive than transporting gaseous hydrogen. This is because of the additional cost and energy required for liquefaction and regasification processes for delivering liquid hydrogen to the market.

6.2.2 Transportation of Hydrogen by Pipeline

There are about 5000 km (about 3125 miles) of hydrogen pipelines worldwide, out of which about 2600 km (about 1625 miles) are in the United States that are mainly used to deliver hydrogen to chemical and refinery facilities (Shell, 2017). Unlike the other two modes (road and ship) of transporting hydrogen, pipelines are used for transporting only gaseous hydrogen. Pipeline appears to be the way to go in transporting large volumes of hydrogen to residential and industrial users of hydrogen. This will be even more so if existing natural gas pipelines can be utilized in convening hydrogen to its various markets. This approach to consider use of existing natural gas pipelines for hydrogen is strengthened by the high compressibility factor and much lower pressure drop of hydrogen in a pipeline over a long distance than that of natural gas, which will afford the possibility of reducing the inlet pressure required for hydrogen transportation via a pipeline to 2–3 MPa (Włodek et al., 2016). It will be too expensive to build a brand new dedicated pipeline system to transport hydrogen. It can be co-mingled or blended with natural gas and be transported in an existing natural gas pipeline.

About 3 million km of natural gas pipelines are in existence globally. Also, there is an established infrastructure for international liquefied natural gas (LNG) shipping, according to Snam, IGU and BCG (2018) and Speirs et al. (2017). Utilization of some of this existing infrastructure to deliver and use hydrogen could significantly boost hydrogen development. However, blending hydrogen with natural gas has several challenges, which include the following, as listed in the IEA (2019) report:

- The energy density of hydrogen is about one-third that of natural gas which reduces the energy content of the delivered blended gas. For example, a 3 percent hydrogen blend in a natural gas transportation pipeline would reduce the energy that the pipeline delivers by approximately 2 percent. Consequently, higher gas volumes would be required to meet a given energy demand by the end user.
- Hydrogen burns much faster than methane. Hence, flame flammability propensity and its associated risk and safety issues are enhanced. New flame detectors would probably be needed for high-blending ratios.
- Variability in the volume of hydrogen blended into the natural gas stream would adversely impact the design and operation of combustion equipment (Abbott et al., 2012]). This can adversely affect the emission of combustion-generated pollutants.
- The upper limit for hydrogen blending in the grid depends on the equipment associated with it.

Alternative to blending will be an integrated approach, which involves insertion of a specialized hydrogen pipeline inside an existing natural gas pipeline – a phenomenon known as pipe-in-pipe approach (Yang et al., 2023).

It is worth mentioning that hydrogen has the potential to cause pipeline degradation of heat-affected zones (HAZ) and fatigue crack growth (FCP) in the base material of the pipeline (Holbrook et al., 2012), consequently leading to pipeline failure. This may be partly because of the relatively lower heating value of hydrogen, and consequently, its (hydrogen) less energy efficiency than that of natural gas (Yang et al., 2023). Another pipeline material issue with hydrogen pipeline is hydrogen embrittlement, which can cause pipeline to become brittle and eventually cause pipeline failure. Research and development efforts are, however, being made to address these issues (Cerniauskas et al., 2020). For example, it has been shown that the presence of small amount (about 150 ppm) of oxygen in the hydrogen is effective in reducing hydrogen embrittlement in metallic materials (Holbrook et al., 2012; Kussmaul, 1998; Michler et al., 2012). Other preventive actions include coating the inside of the pipeline with appropriate specialty coating materials. Upon delivery of the hydrogen at the end use point, it may be necessary to subject the hydrogen to further purification. The level of purity depends on what the hydrogen is used for.

6.2.3 Transportation of Hydrogen via Shipping

As the hydrogen economy becomes more and more global, the importance of hydrogen transportation via ship cannot be overemphasized. Apart from being the only mode of transporting hydrogen in all forms and states (gas, liquid, and solid), it is the only mode of transportation that can provide hydrogen access to the international market. According to Yang et al. (2023), shipping hydrogen appears to be more expensive than transporting it in new and retrofitted pipelines except for when the delivery distance is greater than 700 km when shipping hydrogen becomes cheaper than in new pipelines. However, hydrogen shipping cost appears to be less sensitive to distance when compared with transporting it via pipelines. As mentioned, hydrogen can be shipped in various forms – compressed gaseous, liquid, and in chemical-based forms (ammonia-based and liquid organic carrier-based).

6.2.3.1 Shipping of Compressed Gaseous Hydrogen

Transporting compressed gas by ocean-going vessel is a relatively simple process which involves the following three basic steps:

- Compressing the gas from 2 to 25 MPa of pressure, which requires only about 1.1 kWh/kg of hydrogen according to ASX Announcement (2021), who developed the first compressed hydrogen cargo ship (PV Magazine Australia, 2021).
- Loading the ship with compressed gas containers/vessels. The compressed gas can be injected directly into the ship.
- Off load the ship at the destination.

Although shipping compressed hydrogen appears to be relatively simple that does not require a complex infrastructure, the advantage is negated by the low volumetric storage density (Yang et al., 2023). The first technology demonstration project to build a large, compressed hydrogen cargo ship with two 20 m diameter cylindrical tanks transporting 2000 metric tons of compressed hydrogen was scheduled to be completed by 2023 (Yang et al., 2023). Although transporting compressed gas via ship is restricted to short distance due to its low volume density, it, however, does not require pre- or post-shipment conversion.

6.2.3.2 Shipping of Liquid Hydrogen

In recent years, interest in transporting liquid hydrogen has been increasingly growing. Delivery of hydrogen in its liquid state by ship is like that of LNG. Technology for shipping LNG is mature with a long history and it is being successfully used to transport natural gas under cryogenic conditions across the ocean all over the world (Kimura et al., 2022; Yang et al., 2023). For example, there were about 37 LNG terminals in Europe as of 2021 and about 27 additional planned ones (Yang et al., 2023). Hence, the existing LNG facilities can be retrofitted for importing and/or exporting liquid hydrogen (Seehäfen in der Energiewande: Wasserstoff, 2021). Some liquid hydrogen shipping projects have also been launched in the last few years. For example, a liquid hydrogen ship, which was built by Suiso Frontier in 2020 (Kimura et al., 2022) hauled its first liquid hydrogen consignment (about 75 metric tons at the beginning of 2022 over about 900 km (about 563 miles) according to Werner (2021).

Basically, LH shipment system involves the following steps:

- Liquefaction – This step is where the hydrogen is cooled to about –253°C at high pressure. About 6–10 KWH of energy is required to liquefy a kilogram of hydrogen.
- Loading liquefied hydrogen (LH) on to the ship.
- Shipping LH (kept at –253°C) to the destination.
- Off-loading of LH from the ship.
- Regasification of LH at the destination terminal.

Like LNG, shipping of liquid hydrogen also experiences the boil-off loss issue. Although the boiled-off hydrogen can be re-gasified on board as it is done with LNG, this approach will come with additional cost of providing the extra space needed for the tanks required for storing the excess boil-off, and a complete compressor system for compressing the hydrogen tank will be needed on the ship, all of which will add to the cost and complexity (Yang et al., 2023). Consequently, the tank pressure may rise above the tank limit, thereby necessitating venting the boiled-off hydrogen vapor to the atmosphere (Yang et al., 2023). According to Ortiz-Cebolla et al. (2021), the boil-off impacts limit the maximum sailing time, and consequently, the maximum distance to which liquid hydrogen can be shipped should be in the range of 2500–16,000 km (1563–10,000 miles). It is also recommended that future tanks should be designed to have higher pressure limits and/or equipped with active liquid hydrogen cooling systems (Yang et al., 2023).

While it appears to be cheaper to transport liquid hydrogen than to transport compressed gaseous hydrogen, that advantage could be negated by the high cost incurred during the liquefaction and regasification processes, especially if the transportation of liquid hydrogen is over a short or medium distance. In addition, there is the propensity for a boil-off loss problem to occur during a long-distance delivery, which, consequently, may add to the overall transportation cost, just like the case with compressed gaseous hydrogen, as mentioned earlier.

6.2.3.3 Shipping of Ammonia-Based Hydrogen

In addition to transporting hydrogen in gaseous tube trailers, liquid tankers, and pipelines, hydrogen can also be combined or reacted with other chemicals to form another hydrogen-containing compound, which can then be carried or delivered to an end user where the hydrogen will be separated and used. As mentioned, ammonia is one of such hydrogen-containing chemical compounds in which hydrogen can be carried.

In general, delivery of hydrogen in the form of ammonia involves the following basic steps:

- Ammonia synthesis
- Loading ammonia on to the ship
- Off-loading ammonia from the ship
- Cracking or decomposition of ammonia to release hydrogen

Ammonia synthesis – Industrial production of ammonia has been known for many years worldwide to be through the Haber–Bosch process The process involves the reaction of gaseous hydrogen with molecular nitrogen (N_2) in the presence of catalyst (such as Fe) at high temperatures (in the range of 300–500°C) and pressures of about 20–35 MPa (Klerke et al., 2008; Thomas and Parks, 2006)). This reaction can be represented by the following stoichiometric chemical equation:

$$N_2 + 3H_2 \leftrightarrow 2NH_3 \quad \Delta H^\circ = -91.8 \text{ kJ/mol} \qquad [6.1]$$

In practice, a hydrogen production system/facility (such as steam methane reforming) would be integrated with an ammonia synthesis process system (such as the Haber–Bosch process). Liu (2013) estimated the overall energy consumption for this to be about 8–9 kWh/kg ammonia.

Ammonia cracking or decomposition – This is a process by which hydrogen is recovered from ammonia. It is a stepwise sequence reaction that decomposes ammonia and releases hydrogen. The reaction, which is simply the reverse of the synthesis reaction (Equation 6.1), can be represented by Equation [6.2] (Thomas and Parks, 2006):

$$NH_3\,(g) \rightarrow 0.5N_2\,(g) + 1.5H_2\,(g) \quad \Delta H = +46 \text{ kJ/mol} \qquad [6.2]$$

The ammonia decomposition reaction, governed by Equation [6.2], is slightly endothermic, requiring addition of external heat energy to be sustained. While reaction kinetics demands that high reaction temperatures are required for efficient cracking of ammonia, reasonably high ammonia decomposition can be achieved at low temperatures thermodynamically (Gosnell, 2005; Thomas and Parks, 2006). Ammonia begins to decompose at temperature as low as 200°C. Thermodynamically, 98–99 percent conversion of ammonia to hydrogen is achievable at temperatures as low as 425°C (Thomas and Parks, 2006). An ammonia decomposition reaction is carried out in the presence of a catalyst (such as supported nickel and Ru/S_iO_2 catalysts), and the temperature required for efficient cracking depends on the type of catalyst used. About 6 kWh of energy is required to release 1 kg of hydrogen from ammonia during its decomposition reaction (Andersson and Grönkvist, 2019).

Shipping hydrogen in the form of ammonia is simpler and less complex than hauling compressed gaseous hydrogen which requires high-pressure tanks to hold and transport it, or for transporting hydrogen in liquid form which needs to be liquified, maintained under cryogenic conditions during shipment, and re-gasified at its destination. Ammonia can be shipped at –33°C and atmospheric pressure compared with liquid hydrogen which is shipped at –253°C (Yang et al., 2023). However, ammonia synthesis process is more energy-intensive than hydrogen liquefaction, and ammonia needs to be subjected to a cracking process (which requires additional process step that may consume about 20 percent of hydrogen, according to Yang et al., 2023) to release the hydrogen at the end-use destination.

Shipping of ammonia is a well-developed technology with a relatively long history, reported to be over 100 years by Yang et al. (2023). Annual global shipment of ammonia across the sea is estimated to be about 17.5 million tons, utilizing about 170 ships, 38 export terminals, and 88 import terminals (Laval et al., 2020). Hence, the existing ammonia delivery system can be integrated into a system for carrying hydrogen in the form of ammonia. It appears that interest in developing a commercial-scale ammonia-based hydrogen supply chain is currently at its infancy (ASX Announcement, 2021). In summary, other advantages and disadvantages of transporting hydrogen in ammonia include

- Advantages
  - Relatively more convenient and safe storage conditions during shipping
  - Relatively more convenient handling conditions during shipping
  - Ammonia is versatile – can be used directly as fuel or as feedstock for other industrial processes
  - Low initial shipping and infrastructure costs
- Disadvantages
  - Most expensive in general
  - Hazardous properties of chemical carrier

6.2.3.4 Shipping of LOHC-Based Hydrogen

In addition to using ammonia to carry hydrogen, other chemicals can be bonded with hydrogen and then be used to carry hydrogen from its point of production to the point of consumption or utilization. Liquid-organic hydrogen carriers (LOHCs) are liquid or semi-liquid-organic substances at ambient temperature and contain unsaturated bonds that allow them to be hydrogenated when in contact with hydrogen molecules (Yang et al., 2023).

LOHC compounds can be divided into two main categories – homocyclic and heterocyclic compounds (Yang et al., 2023). Examples of homocyclic LOHC include homocyclic methylcyclohexane (MCH) (also known as toluene), decalin/naphthalene, and some multicomponent systems, such as the one containing perhydro-benzyl toluene (H12-BT)/benzyl toluene (H0-BT) and perhydro-dibenzyl toluene (H18-DBT)/dibenzyl toluene (H0-DBT) (Preuster, 2021; Yang et al., 2023). Examples of heterocyclic LOHCs, which are more favorable over homocyclic LOHCs because of their positive attributes (low dehydrogenation temperature, reversibility, favorable thermodynamics, and kinetics), include N-methyl per-hydroindole/N-methylindole,

2-methyl perhydro-indole/2-methylindole, 1,2-dimethyl perhydroindole/1,2-dimethyl indole (Yang et al., 2023). LOHC-based hydrogen can be shipped and handled more safely and easily than liquid hydrogen and higher-LOHC-based hydrogen can be carried per unit volume of space than compressed gaseous hydrogen (Seehafen, 2021), thereby requiring less space per unit volume of hydrogen transported.

Basically, LOHC-based hydrogen delivery value chain involves the following steps:

- Hydrogenation of LOHC
- Storage of hydrogen-rich LOHC
- Loading ship with hydrogen-rich LOHC
- Shipping and off-loading hydrogen-rich LOHC
- Storage of hydrogen-rich LOHC (if necessary)
- Dehydrogenation of hydrogen-rich LOHC

Finally, some purification of the hydrogen may be necessary depending on the purpose the hydrogen will be used for at the consumer end. Hydrogenation is a chemical reaction between molecular hydrogen and another compound or element, usually in the presence of a catalyst such as nickel, palladium, or platinum. The process is commonly employed to reduce or saturate organic compounds. Hence, hydrogenation simply involves the conversion of unsaturated liquid hydrocarbon compound to saturated liquid hydrocarbon compound by reacting it with hydrogen in the presence of a metal catalyst. Liquid organic compounds listed earlier are the unsaturated compounds in this case. Hydrogenation reaction is an exothermic reaction (i.e., it gives off heat), while dehydrogenation reaction is an endothermic reaction (i.e., requires addition of energy (about 9 kWh/kgH_2) to be sustained). Hydrogenation and dehydrogenation reaction temperatures and pressures vary with different liquid organic compounds. More details can be found in a recent review paper by Yang et al. (2023). There is no boil-off issue during shipment of LOHC-based hydrogen, thereby making it safer to transport.

According to Yang et al. (2023), it is worth noting that commercial deployment of LOHC shipping system is in the testing stage. There is still need for more research and development (R&D) to address the energy issue with dehydrogenation. It is also worth noting that long-distance transportation and local distribution of hydrogen is difficult because of its low energy density. Current commercial production of toluene (one of the most used LOHC chemicals) is estimated to be about 22 million tons yearly, which could carry 1.4 million tons of hydrogen if it were to be used as an LOHC (IEA, 2019). Compression, liquefaction, or hydrogenation are possible options to address this issue. However, there are advantages and disadvantages associated with each option, as enumerated earlier in Section 6.2.3. The economic competitiveness will vary with geographical location, distance, scale, and end-use requirements. Also, the assessment by Niermann et al. (2021) shows that long-distance transport favors LOHC, while short-distance transport via pipelines can be used for lower costs, and that no specific transportation chain is superior to all systems under all circumstances.

The merits and drawbacks of transporting hydrogen in LOHC can be summarized as follows:

- Advantages
  - Relatively more convenient and safe storage conditions during shipping
  - Relatively more convenient handling conditions during shipping
  - Can be transported as liquids without the need for cooling
  - Low initial shipping and infrastructure costs
- Drawbacks
  - Hazardous properties of chemical carrier
  - Energy-intensive dehydrogenation process

6.2.4 Local Distribution of Hydrogen

Upon arrival at the import terminal or transmission hub and after being released from its appropriate chemical carrier (such as ammonia, LOHC), hydrogen can then be delivered to its final consumers through a local distribution network. The appropriate methods for local distribution of hydrogen are mainly by road trucks and local pipelines. Like the long-distance transportation mode, choice of best option for local distribution of hydrogen will be determined by factors such as volume to be distributed or delivered, delivery distance, and purpose for which the hydrogen will be used for by the end user.

6.2.4.1 Trucks

For distances less than 300 km (about 188 miles), compressed gas trailer trucks are currently the method of choice used to locally distribute hydrogen, while liquid hydrogen tanker trucks are preferred in cases where there is reliable demand, and the liquefaction costs can be offset by the lower unit costs of hydrogen transport (IEA, 2019). The hydrogen is delivered via trailers loaded with tubes in both methods.

While a single trailer can haul and hold up to 1,100 kg (theoretically) of compressed hydrogen in lightweight composite cylinders (at 500 bar), up to 4000 kg of liquefied hydrogen can be hauled by well-insulated cryogenic tanker trucks (IEA, 2019). As also indicated in the IEA report (2019), about 5000 kg of hydrogen and 1700 kg of hydrogen in the form of ammonia and in the form of LOHC, respectively, could be hauled in a road tanker.

6.2.4.2 Pipelines

Local distribution of hydrogen through pipelines is possible. It could either be done using the existing low-pressure network of pipelines made of polyethylene or fiber-reinforced polymer or through new pipelines dedicated to hydrogen. Distribution pipelines for natural gas are extensive in residential areas with high demand for heating and cooking in many parts of the world such as northern Europe, the People's Republic of China, and North America, reaching into urban areas as well as industrial clusters (IEA, 2019). These extensive existing natural gas distribution network of pipelines can be retrofitted for hydrogen distribution.

New dedicated hydrogen distribution pipelines would be very expensive, especially on the scale required for supplying hydrogen to residential sector (IEA, 2019).

References

Abbott, D.J., Bowers, J.P., and James, S.R. (2012). The Impact of Natural Gas Composition Variations on the Operation of Gas Turbines for Power Generation. The Future of Gas Turbine Technology 6th International Conference, 17–18 October 2012, Brussels, Belgium. https://gasgov-mst-files.s3.euwest-1.amazonaws.com/s3fs-public/ggf/Impact%20of%20Natural%20Gas%20Composition%20- %20Paper_0.pdf.

Andersson, J. and Grönkvist, S. (2019). Large-Scale Storage of Hydrogen. *Int. J. Hydrogen Energy*, 44, 11901–11919.

ASX Announcement. (2021). GEV Scoping Study Delivers Zero Emission Supply Chain for Green Hydrogen. Global Energy Ventures Ltd. https://buyhydrogen.com.au/wp-content/ uploads/2021/04/H2-Supply-Chain-Scoping-Study-IssuedGEV-210301.pdf (28 February 2023, last accessed).

Ball, M. and Weeda, M. (2016). The Hydrogen Economy—Vision or Reality? In *Compendium of Hydrogen Energy,* Ball, M., Basile, A., Veziroğlu, T.N. (Eds.), pp. 237–266. Oxford, UK: Woodhead Publishing.

Cerniauskas, S., Junco, A.J.C., Grube, T., et al. (2020). Options of Natural Gas Pipeline Reassignment for Hydrogen: Cost Assessment for a Germany Case Study. *Int. J. Hydrogen Energy*, 45, 12095–12107.

Galassi, M.C., Baraldi D., Iborra B.A., et al. (2012). CFD Analysis of Fast Filling Scenarios for 70 MPa Hydrogen Type IV Tanks. *Int. J. Hydrogen Energy*, 37, 6886–6892.

Gosnell, J. (2005). KBR Hydrogen Conference, Argonne National Laboratory, October 13. www.energy.iastate.edu/renewable/biomass/download/2005/Gosnell_production.pdf.

Hexagon-Purus-X-STORE-Gas. (2022). Container-Modules-VersionADR-V2-Full-Carbon-Design-500Bar-H2-1.pdf. www.californiahydrogen.org/wp-content/uploads/2021/06/ Hexagon-Purus-X-STORE-Gas_Container-Modules-VersionADR-V2-Full-Carbon-Design-500Bar-H2-1.pdf (19 September 2022, last accessed).

Holbrook, J.H., Cialone, H.J., Collings, E.W., et al. (2012). Control of Hydrogen Embrittlement of Metals by Chemical Inhibitors and Coatings. In *Gaseous Hydrogen Embrittlement of Materials in Energy Technologies. Volume 2: Mechanisms, Modelling and Future Developments*, Gangloff, R.P., and Somerday, B.P. (Eds.), pp. 129–153. Sawston, UK: Woodhead Publishing.

International Energy Agency. (2019). The Future of Hydrogen – Seizing Today's Opportunities. A Report by the IEA for the G20.

Kimura, S., Kutani, I., Ikeda, O., et al. (2022). Demand and Supply Potential of Hydrogen Energy in East Asia: Phase 2. Research Project Report 16. ERIA, 2020. www.eria.org/publications/demand-andsupply-potential-of-hydrogen-energy-in-east-asia-phase-2/.

Klerke A., Christensen, C.H., Nørskov J.K., et al. (2008). Ammonia for Hydrogen Storage: Challenges and Opportunities. *J. Mater. Chem.*, 18, 2304.

Kussmaul, K. (1998). Fracture Mechanical Behavior of the Steel 15 MnNi 6 3 in Argon and in High Pressure Hydrogen Gas with Admixtures of Oxygen. *Int. J. Hydrogen Energy*, 23, 577–582.

Laval, A., Hafnia, Topsøe, H., and Gamesa, V.S. (2020). Ammonia Fuel: An Industrial View of Ammonia as a Marine Fuel. White Paper. https://hafniabw.com/wp-content/ uploads/2020/08/Ammonfuel-Report-an-industrial-view-ofammonia-as-a-marine-fuel.pdf.

Liu, H. (2013). *Ammonia Synthesis Catalysts: Innovation and Practice*. Singapore: World Scientific Publishing, Co. Pte. Ltd.

Michler, T., Boitsov, I.E., Malkov, I.L., et al. (2012). Assessing the Effect of Low Oxygen Concentrations in Gaseous Hydrogen Embrittlement of DIN 1.4301 and 1.1200 Seels at High Gas Pressures. *Corros. Sci.*, 65, 169–177.

Niermann, M., S. Timmerberg, S., Drünert, S., and Kaltschmitt, M. (2021). Liquid Organic Hydrogen Carriers and Alternatives for International Transport of Renewable Hydrogen. *Renewable Sustain. Energy Rev.*, 135, 110171.

Ortiz-Cebolla, R., Dolci, F., Weidner, E. (2021). Assessment of Hydrogen Delivery Options. The European Commission's Science and Knowledge Service, Joint Research Centre. https://ec.europa.eu/jrc/sites/default/files/jrc124206_assessment_of_hydrogen_delivery_ options.pdf.

Preuster, P. (2021). Die LOHC-Technologie: Flüssige Organische Wasserstoffträger'. Presented at the invited talk, 22 June.

PV Magazine Australia. (2021). GEV To Construct 430 Tons Hydrogen Cargo Ship for Small Scale Grid Blending Market. www.pv-magazineaustralia.com/2021/06/08/gev-to-construct-430t-hydrogencargo-ship-for-small-scale-grid-blending-market/.

Reuß, M., Dimos, P., Léon, A., et al. (2021). Hydrogen Road Transport Analysis in the Energy System: A Case Study for Germany Through 2050. *Energies*, 14, 3166.

Seehäfen in der Energiewande: Wasserstoff. (2021). Hamburg: Zentralverband der deutschen Seehafenbetriebe e.V. (ZDS). https://zds-seehaefen.de/wp-content/uploads/2021/06/2021-06-03_ZDS_Wasserstoff_ Arbeitspapier_Juni21.pdf.

Shell. (2017). Shell Hydrogen Study: Energy of the Future. www.shell.de/medien/shellpublikationen/shell-hydrogen-study/_jcr_content/par/toptasks_e705.stream/ 1497968967778/1c581c203c88bea74d07c3e3855cf8a4f90d587e/shell-hydrogen-study.pdf.

Snam, IGU (International Gas Union) and BCG (Boston Consulting Group). (2018). Global Gas Report 2018. www.snam.it/export/sites/snam-rp/repository/file/gas_naturale/global-gasreport/global_gas_report_2018.pdf.

Speirs, J., Balcombe, P., Johnson E., Martin, J., Brandon, N. and Hawkes, A.. (2017). A Greener Gas Grid: What Are the Options? Sustainable Gas Institute, Imperial College London. www.sustainablegasinstitute.org/wp-content/uploads/2017/12/SGI-A-greener gas-grid-what-are-the-options-WP3.pdf?noredirect=1.

Thomas, G. and Parks, G. (2006). U.S. Department of Energy. Potential Roles of Ammonia in a Hydrogen Economy A Study of Issues Related to the Use Ammonia for On-Board Vehicular Hydrogen Storage. A Study Report for the U.S. Department of Energy.

U.S. Department of Energy. (2020). Department of Energy Hydrogen Program Plan. DOE/EE – 2128.

U.S. Department of Energy, Office of Hydrogen and Fuel Cell Technology. (2022). Hydrogen Tube Trailers. www.energy.gov/eere/fuelcells/hydrogen-tube-trailers.

Werner, P. (2021). Kawasaki Stellt ersten Wasserstofftanker fertig. www.golem.de/news/wasserstoff-kawasaki-stelltersten-wasserstofftanker-fertig-2105-156880.html.

Włodek, T., Łaciak, M., Kurowska, K., et al. (2016). Thermodynamic Analysis of Hydrogen Pipeline Transportation: Selected Aspects. *AGH Drilling, Oil, Gas*, 33, 379–396.

Yang, M., Hunger, R., Berrettoni, S., Sprecher, B., and Wang, B. (2023). A Review of Hydrogen Storage and Transport Technologies. *Clean Energy*, 7(1), 190–216.

7

Hydrogen Storage

7.1 Introduction

Hydrogen, an industrial gas, can be stored either as a compressed or as a refrigerated liquefied gas. Since the beginning of the 20th century, hydrogen is stored in seamless steel cylinders. In the late 1960s, tubes made of seamless steel were also used, with specific attention paid to hydrogen embrittlement in the 1970s. Aluminum cylinders were also used for hydrogen storage till the end of the 1960s but were found to be cost-prohibitive and they have smaller water capacity compared with steel cylinders. Metallic cylinders can be hoop-wrapped to further increase the working pressure of hydrogen tanks or to slightly reduce the weight. Development of fully wrapped tanks was begun in the 1980s for space or military applications. Their use for portable applications such as on-board storage of natural gas for vehicles began thereafter because of their light weight.

The importance of storage in the hydrogen value chain cannot be over-emphasized as the methods of transporting hydrogen and its utilization are determined by the way in which it is stored. Hydrogen storage is required at various steps of the value chain – at production site, during transportation, such as shipping of liquid hydrogen, ammonia-based hydrogen, and liquid organic hydrogen carrier, at import terminals prior to its local distribution to various end users. This chapter examines the various technologies for hydrogen storage.

Typically, compressed gaseous hydrogen or liquid hydrogen is stored in tanks for small-scale mobile or stationary use. But as national and international demand for hydrogen grows, a much broader variety of storage options will be needed. For example, hydrogen storage will be needed at the export terminal (prior to shipping), during shipping, and at the import terminal prior to being distributed for local consumption, and as well at vehicle refueling and city gates (IEA, 2019). If hydrogen breaks into a larger demand market such as in electric power industry and other industrial sector, relatively much larger storage facilities would be required to meet and sustain the long-term demand (IEA, 2019). The choice of the most suitable storage technology potion depends on some factors, which include the quantity of hydrogen to be stored, storage duration, the required speed of discharge, and the availability and flexibility of various storage options (IEA, 2019). Geological storage is generally the most suitable option for large-scale and long-term storage, while tanks are more suitable for short-term and small-scale storage.

DOI: 10.1201/9781003348283-7

In general, the storage technologies that are currently available for storing hydrogen are those that stem from the chemical and natural gas industries. Gaseous and liquid hydrogen can be physically stored in storage tanks both over ground and underground, as well as in geological formation. Hydrogen can be stored by chemically or physically combining it with appropriate liquid or solid materials. As discussed in Chapter 6, hydrogen can be chemically bonded/reacted/combined with some chemical compounds or elements to form stable compounds in which the hydrogen can be stored. For example, hydrogen can be reacted with molecular nitrogen to form ammonia in which hydrogen is stored. Similarly, some liquid organic chemicals can be combined or reacted with hydrogen to form liquid organic hydrogen carriers in which hydrogen is stored and carried, as discussed in Chapter 6. As well, some intermetallic compounds such as titanic-based alloys can also be physically or chemically combined with hydrogen and used to store hydrogen. The different hydrogen storage technology options, as well as the energy required to store and to release the hydrogen when needed for use, will be discussed in Section 7.2 of this chapter.

7.2 Hydrogen Storage Technologies

A hydrogen tank is a canister used to store compressed hydrogen. Hydrogen can be stored in gaseous and liquid forms. A gaseous storage tank works by storing compressed hydrogen under pressure, while a cryogenic storage tank works by storing hydrogen in liquid form. Technology options for storing hydrogen include (1) use of storage tanks or pressure vessels, (2) underground storage in geological formation, (3) cryo-compressed storage, (4), storage in liquid form, (5) storage in chemical carriers such as ammonia and liquid organic hydrocarbon carrier (LOHC) and ammonia, (6) storage in metallic materials such as metal hydrides, (7) storage in fuels such as synthetic fuels. According to Dutta (2014) and Yang et al. (2023), hydrogen storage system can be technically characterized using five main factors – gravimetric density, volumetric density, operating temperature, number of cycles, and the rate of filing the storage system. These characteristics vary with what hydrogen is used for and with region or country where the hydrogen is used.

Generally, hydrogen is stored by changing its physical state, either by compressing or liquifying it (known as physical-based storage) or by chemically or physically combining hydrogen with another molecule or element (such as ammonia and a variety of LOHCs described in Chapter 6) in which the hydrogen is stored (a phenomenon known as material-based storage). The physical-based storage techniques include compressed gaseous hydrogen storage, cryo-compressed hydrogen storage, and liquid hydrogen storage, while the material-based storage methods are hydrogen stored in LOHCs, metal hydrides, and fuels/power. Although the material-based storage methods provide a safer and more storage density than the physical-based storage systems, the material-based technologies are, however, still in the laboratory and demonstration stages (Yang et al., 2023).

7.2.1 Hydrogen Storage in Compressed Tanks (Pressure Vessels)

Compressed gaseous hydrogen storage is simply, as defined by Hua et al. (2010), storing hydrogen in pressure vessels at high pressures, such as in the range of 350–700 bar (approximately 5000–10,000 psi). Like petroleum products, such as gasoline, hydrogen can be stored before and after being transported to its end use point. However, it will take much more space to store a unit quantity of hydrogen compared with storing the same quantity of gasoline because the energy density of hydrogen is relatively lower than that of gasoline. For example, compressed hydrogen (at a pressure of 70 MPa) with only 15 percent of the energy density of gasoline would require almost seven times the space to store the same amount of gasoline (IEA, 2019). Contrarily, compressed hydrogen tanks have a higher energy density than lithium-ion batteries, and so consequently less storage space is required for hydrogen in hydrogen-fueled cars or trucks than for lithium-ion battery space in electric vehicles (IEA, 2019).

Compressed gas tanks/pressure vessels provide the easiest or most convenient method to store gaseous hydrogen. Compressed or liquefied hydrogen storage tanks, with high efficiencies and discharge rates, are suitable for storing hydrogen for small-scale local distribution. Basically, all it involves is compressing the gas and filling it into the pressure vessels or tanks. However, the tanks or vessels must have the appropriate characteristics/properties and specifications, which vary with the quantity of gas, the duration of storage, and the application for which the gas will be used for, among other requirements. Presently, there are four types of pressure vessels being used for storing hydrogen – Types I through IV, which have been briefly described in Chapter 6. A more detailed description of the various relevant types of pressure vessels for hydrogen storage is provided further.

Type I pressure vessels for storing hydrogen, which is made of metallic materials such as aluminum and steel, is designed to withstand and operate at pressures ranging from 15–30 MPa (Barthelemy et al., 2017). The application of Type I storage tank is limited to on-site storage of hydrogen as industrial gas due to its low storage density. The middle portion of the cylindrical vessel can be hoop-wrapped with resin-impregnated fiber to strengthen the vessel and enable it to store hydrogen at a higher pressure, thereby transitioning it into Type II pressure vessel (Yang et al., 2023). As mentioned, Type II pressure vessel is a Type I vessel hoop-wrapped with glass fiber composite designed to operate at and withstand a wider range of pressure (10–90 MPa) (Barthelemy et al., 2017), with a maximum tolerant pressure (which is the highest of all the four types of pressure vessels) of about 100 MPa (Yang et al., 2023). According to Parks et al. (2014) and as stated by Yang et al. (2023), Type II pressure tanks are typically used for stationary high-pressure gas storage, such as cascade hydrogen storage at a hydrogen re-fueling station with 87.5 MPa. Type III pressure vessel is made with full composite wrap with metal liner designed for a working pressure range of 30–70 MPa and used for on-board hydrogen storage.

It is worth mentioning and as noted by Yang et al. (2023) in their recent review paper that the seamless metal cylinder and liner for Type I, II, and III pressure vessels are manufactured in a very similar way – the incoming metal slugs or plates are deep-drawn into the shell, which are subsequently stamped and hot-formed to form a neck. Type IV vessels are made of fully composite such as high-density polyethylene inner with glass or carbon fiber. The glass or carbon fibers provide additional strength to the vessel liners. In addition, the vessel can be hoop-wrapped, polar-wrapped, or helical-wrapped thereon, after which the fibers are protected by applying cured resins (Yang et al., 2023).

7.2.2 Hydrogen Storage in Geological Formation (Underground Storage)

In addition to pressure vessels or tanks, there are other possible options for large-scale and long-term hydrogen storage. These other options include depleted oil and natural gas reservoirs, salt caverns, and aquifers (HyUnder, 2014; IEA, 2019). These alternative storage options are currently engaged in natural gas storage with high efficiency – injected hydrogen/extracted hydrogen ratio – and low costs, and consequently offer promise for cheap hydrogen storage options (Bünger et al., 2014; Perera, 2023).

The chemical industry in the United Kingdom and the United States has been utilizing salt caverns to efficiently and cost-effectively store hydrogen with little or no risk of contamination since the 1970s and the 1980s (IEA, 2019; Bünger et al., 2014). The largest salt cavern hydrogen storage facility currently exists in the United States with an estimated storage capacity of between 20,000 and 30,000 metric tons. There are three salt cavern hydrogen storage systems in the United Kingdom with estimated storage capacity of about 1000 metric tons, and a 3500 metric tons salt cavern storage demonstration project is expected to have been completed in 2023 in Germany (IEA, 2019). More details about these underground hydrogen storage projects as well as some others across the world can be found elsewhere (Yang et al., 2023; Zivar et al., 2021).

Of all the underground hydrogen storage options, the salt cavern appears to be the best option because of its unique favorable attributes – tightness of its deposits, mechanical properties, and its resistance to chemical reactions (Tarkowski, 2019), and also its sealing ability provided by the viscoelastic evaporitic rocks of the salt cavern, and its resistance microbial consumption of stored hydrogen due to the saline environment in the salt cavern system (Sainz-Garcia et al., 2017).

Although depleted oil and gas reservoirs possess relatively larger storage capacity than salt caverns and with well-known and well-characterized geological structure (Tarkowski, 2019), they, however, contain contaminants (which may mix with hydrogen) that would need to be removed before the hydrogen is used for some applications (such as fuel cells), and they are permeable, causing unwanted leakage problem (IEA, 2019).

Although water aquifers have been reported for storing town gas (containing 50–60 percent of hydrogen) for years, their suitability for pure hydrogen storage is yet to be proven. As well, some effects of some physicochemical processes (such as wettability, interfacial tension, diffusion, adsorption, and solubility) on the rock–fluid interactions – geochemical interactions and microbial reactions – have been reported during underground hydrogen storage in depleted gas reservoirs (Perera, 2023).

Also, the suitability and the techno-economic feasibility of utilizing depleted reservoirs and aquifers for commercial storage of pure hydrogen is also yet to be demonstrated. Other drawbacks of using aquifers for underground hydrogen storage include the propensity for leakage along undetected faults because of the porous nature of the aquifer and biochemical reactions (Yang et al., 2023; Perera, 2023). There is also the potential for the minerals in the reservoir rock to react with hydrogen under the reservoir conditions (<130°C and 35 MPa), consequently leading to dissolution and precipitation of the minerals (Perera, 2023). They, however, hold promise for seasonal hydrogen storage, especially with no other alternative hydrogen storage options such as salt caverns (IEA, 2019). According to the IEA (2019) report, the high-pressure and high discharge rate attributes of geological storage systems offer them a very good promise for long-term and large-scale storage of hydrogen; they are, however, less suitable for short-term and small-scale storage of hydrogen because of their geographical distribution, large size, and minimum pressure requirements.

7.2.3 Liquid Hydrogen Storage

Changing hydrogen from gaseous state to liquid state is another way of storing it. As mentioned, this is achieved through a process known as liquefaction. This simply involves cooling of hydrogen to a temperature of about –253°C at a pressure of 101 KPa (1 atmosphere). While the liquefaction of hydrogen gas increases the gravimetric and volumetric density (i.e., its storage ability), the process is, however, relatively more complex and energy consuming than hydrogen compression. These characteristics are attributed by Valenti (2016), and Yang et al. (2023) to the following:

- its (hydrogen) low evaporation temperature and critical point (–240°C);
- small molecule size that makes it behave close to the ideal gas at a relatively high temperature; and
- lower parahydrogen enthalpy of vaporization (447 kJ/kg) at –253°C than the enthalpy of exothermic conversion (requiring 532 kJ/kg) from normal to equilibrium hydrogen at the same temperature because of the ortho-to-para conversion of hydrogen.

Although a process for liquefying air has been in existence since the late 19th century during which William Hampton and Carl von Linde independently invented their air liquefaction processes, this process was found not to be suitable for hydrogen liquefaction (Barron, 2000). In 1902, George Claude came out with his own process, which is an improved Linde-Hampton process (Yang et al., 2023). Detailed description of both the Claude-Hampton and Claude processes could be found in a recent review paper by Yang et al. (2023). After liquefaction, the hydrogen can be stored in pressure vessels and storage tankers/tanks.

There are currently two hydrogen liquefaction plants in Leuna, Germany (one has been in operation since 2007, while the other was scheduled to be commissioned in 2021), both of which are based on the Claude liquefaction process equipped with more low-temperature expansion turbines and more heat exchangers and integrated catalytic ortho-to-parahydrogen converters (Yang et al., 2023). A new cooling system, which combines a Ranque–Hilsch vortex tube with endothermic conversion of para-to-orthohydrogen conversion, has been developed by a group of researchers at the University of Washington that can be used instead of the commonly used refrigeration system based on Claude cycle (Matvee and Leachman, 2021; Leachman, 2015). As outlined by Yang et al. (2023), the process of converting para-orthohydrogen in a Ranque–Hilsch vortex tube basically involves pre-cooling of a pressurized stream of hydrogen in a bath of liquid nitrogen at a temperature of −196°C with about 50 percent equilibrium para, and then injected into a vortex tube and then into a swirl chamber, where the hydrogen stream is subjected to a high-speed rotation and gradually separated into two fluids with different temperatures, from which the orthohydrogen-rich stream and the parahydrogen-rich stream are produced. More detailed description of the process can be found in a recent review paper by Yang et al. (2023).

While the Ranque–Hilsch para-orthohydrogen conversion technology has a versatile application to hydrogen liquefaction because of its modularity (Leachman, 2015), it is plagued with the issue of boil-off loss (which can range from 1 to 5 percent for liquid hydrogen storage tanks, according to Derking, 2019) resulting from heat transfer (conduction, convection, and radiation) from the cryogenic liquid hydrogen tanks. Heat absorbed by liquid hydrogen can result in its evaporation. However, several corrective actions to address these thermal issues and maintain the cryogenic tanks well insulated have been reported by Derking (2019).

Yang and co-workers noted in their paper (2023) that several methods have been developed to further prevent boil-off loss of liquid hydrogen tanks, including the use of multilayer insulation technology operated under high vacuum, a technique normally used for storing liquid helium. Swanger et al. (2017) claim that boil-off loss can be completely prevented by integrating a refrigeration system with a multilayer insulated tank.

7.2.4 Cryo-Compressed Hydrogen Storage

Sections 7.2 and 7.3 discuss storage of hydrogen in its gaseous (compressed hydrogen) and liquid state (liquid hydrogen). A discussion of storing hydrogen in its combined state of liquid and vapor – cryo-compressed storage – is presented in this section. According to Ahluwalia et al. (2018), cryo-compressed hydrogen storage basically refers to hydrogen storage at cryogenic temperatures in a vessel that can withstand pressures up to 250–350 atm. As indicated in the assessment report by Ahluwalia et al. (2018), cryo-compressed hydrogen storage can include liquid hydrogen, cold compressed hydrogen, or hydrogen in a two-phase region (saturated liquid and vapor). Cryo-compression is achieved by first subjecting the hydrogen through a series of compression and cooling stages until a desired pressure is obtained. After achieving the desired pressure at the *N*th stage of compression, the compressed hydrogen is sent into a refrigeration process. Also, the number of stages required to compress the hydrogen to a required pressure can be calculated knowing the initial and final required pressures.

According to Yang et al. (2023), cryo-compressed hydrogen can demonstrate a relatively higher storage density than liquid hydrogen at higher temperature, while Kunze and Kircher (2012) reported on their various study efforts that investigated the feasibility of directly producing cryo-compressed hydrogen from liquid hydrogen (rather than from gaseous hydrogen) using a system that combines a high-pressure hydrogen cryopump with a liquefaction/vaporizing unit (Yang et al. 2023). This combined process technology will enable direct dispensation vaporizer potential for a direct cryo-compressed hydrogen into its application unit (such as fuel cell electric vehicle) and easier adaptation to liquid hydrogen-based hydrogen re-fueling stations without the need for an intermediate storage unit (Yang et al., 2023). Other studies on the suitability of cryo-compressed hydrogen storage for on-board application can be found elsewhere (Ahluwalia et al., 2018; Moreno-Blanco et al., 2019; Xu et al., 2020).

Basically, and as mentioned in Chapter 6 and reported by Brunne and Kircher (2016) and Yang et al. (2023), cryo-compressed hydrogen can be stored in Type III compressed gaseous hydrogen pressure vessel with multilayer insulation and enclosed under vacuum.

7.2.5 Chemical-Based Hydrogen Storage

Interest in the use of chemical materials such as ammonia, methanol, and LOHCs in storing hydrogen (material-based storage) has recently gone up. For this section of the book, the focus is, therefore, on these chemical-based hydrogen storage materials. Chemical-based hydrogen storage basically involves the use of chemical materials that allows hydrogen to be bonded onto them, thereby providing hydrogen with some sort of storage, from which the hydrogen can be released when needed. The process by which hydrogen is bonded with the chemicals is referred to as hydrogenation, while the chemical reaction governing the release of the hydrogen thereafter when required is known as dehydrogenation process. The chemicals of interest here are ammonia, methanol, and LOHCs.

7.2.5.1 Liquid Organic Hydrogen Carrier as a Hydrogen Storage

LOHCs have the characteristic that allows hydrogen to be bonded onto them – hydrogenation – thereby providing hydrogen some sort of storage, which can thereafter be released when needed through a process known as dehydrogenation. This section focusses on the use of LOHCs for hydrogen storage. Hydrogenation is typically a catalytic and exothermic reaction, while dehydrogenation is an endothermic catalytic reaction.

As mentioned in Chapter 6, LOHCs are liquid or semi-liquid organic compounds or elements at room temperature containing unsaturated bonds that attract hydrogen molecules when in contact with them. There are some properties required of LOHCs to effectively provide adequate storage system for hydrogen. These, according to Yang et al. (2023) and Rao and Yoon (2020), include the following:

- Safe and non-toxic
- Have relatively higher volumetric and gravimetric storage density (>56 kg/m³ and >6 weight percent, respectively)
- Low melting points (<–30°C) and higher boiling point (>300°C)
- Consumes relatively lower energy during its dehydrogenation reaction
- Possess distinct selectivity for hydrogenation and dehydrogenation
- Amenable to integration or retrofitting with existing fuel infrastructures
- Economic or cost-effective

These criteria are, however, difficult to meet by any of the reportedly several LOHCs developed over the years. Some description of the LOHC-based materials is given in Chapter 6, and a more detailed description can be found elsewhere (Yang et al., 2023).

Basically, there are two main categories of LOHC (homocyclic compounds and heterocyclic compounds) that have got research attention. Toluene (also known as homocyclic methylcyclohexane), decalin or naphthalene, and perhydro-benzyl toluene-containing compounds are good examples of homocyclic LOHC compounds. Heterocyclic LOHCs include N-methyl perhydroindole/N-methylindole, 2-methyl perhydro-indole/2-methylindole, 1,2-dimethyl perhydroindole/1,2-dimethyl indole (Yang et al., 2023). Heterocyclic LOHCs are more favorable over homocyclic LOHOs because of their low dehydrogenation temperature that avoids coke formation, reversibility, favorable thermodynamics, and kinetics. In contrast, dehydrogenation of homocyclic LOHCs (especially toluene) occurs at high temperature (at about 350°C) and consequently results in the formation of coke and other undesirable products. However, Yan and co-workers (2018) in their study found that the problem of coke formation during dehydrogenation of toluene at high temperature could be abated if the dehydrogenation reaction is carried out in the presence of some bimetallic catalyst, such as platinum–tin (Pt–Sn). A summary of selected hydrogenation and dehydrogenation reaction conditions (reaction temperature and catalyst) and hydrogen storage density for LOHC-based can be found in a review paper by Yang et al. (2023).

Briefly, they are, as observed by Yang et al. (2023) as follows:

- The temperature for both the hydrogenation and dehydrogenation reactions varies with LOHC.
- Ru/Al_2O_3 catalyst appears to be adequate for hydrogenation of most of the LOHC, while the catalyst used for the dehydrogenation process seems to vary between Pt/Al_2O_3 and Pd/Al_2O_3 with LOHC.

- The most promising homocyclic LOHC is a multicomponent system that contains perhydro-benzyl toluene (H_{12}-BT)/benzyl toluene (H_0-BT) and perhydro-dibenzyl toluene (H_{18}-DBT)/dibenzyl toluene, with faster hydrogen releasing potential in the presence of Pt/Al_2O_3 catalyst at moderate temperature (about 270°C).

Several studies have been conducted on the fundamentals of LOHC hydrogenation and dehydrogenation including that conducted by Leinweber and Müller (2018, who studied the reaction pathway and kinetics of monobenzyl toluene hydrogenation. It was shown by the results of their study that the hydrogenation reaction occurs predominantly in stepwise hydrogenation of the aromatic ring and that the hydrogenation of monobenzyl toluene possibly involves two reaction pathways – hydrogenation of the mono-substituted side ring (toluene) and di-substituted main ring (xylene).

The benefits of LOHCs are mainly related to their ability to safely and efficiently store and transport hydrogen. These include not being toxic, possessing high purity, and are amenable to existing refinery and transportation infrastructure, and consequently reduces the need for new investments and facilitates the integration of hydrogen into the energy system.

7.2.5.2 Storing Hydrogen in Ammonia

The principal industrial process for producing ammonia is through the Haber–Bosch process which is carried out by reacting molecular nitrogen with hydrogen at 300–500°C and 20–35 MPa pressure in the presence of a metallic catalyst, such as iron (Fe), as alluded to in Chapter 6. The process, which is also known as ammonia synthesis, is represented by the following reversible reaction shown in Equation [7.1] (Klerke et al., 2008):

$$N_2 + 3H_2 \leftrightarrow 2NH_3 \quad \Delta H° = -91.8 \text{ kJ/mol} \qquad [7.1]$$

Basically, the ammonia synthesis process, which is an exothermic reaction, is a hydrogenation process where nitrogen is hydrogenated to form ammonia to provide hydrogen some sort of storage. The formed ammonia can then be liquified at –33°C and atmospheric pressure or at 20°C under a pressure of 0.75 MPa (Yang et al., 2023) and stored appropriately in pressure vessels/storage tanks, depending on the available storage space, demand, and mode of utilization. For industrial storage scale, hydrogen production system/facility (such as steam methane reforming) would be integrated with an ammonia synthesis process plant (such as the Haber–Bosch process). When needed thereafter, the hydrogen can be released from the ammonia by subjecting the ammonia to a dehydrogenation or decomposition process. This can be achieved by reversing Equation [7.1]. The dehydrogenation of ammonia to release hydrogen is represented by Equation [7.2].

$$NH_3\,(g) \rightarrow 0.5N_2\,(g) + 1.5H_2\,(g) \quad \Delta H = +46 \text{ kJ/mol} \qquad [7.2]$$

As mentioned in Chapter 6, the ammonia decomposition reaction, which is represented by Equation [7.2], is conducted in the presence of a catalyst (such as supported nickel and Ru/S_iO_2), and is slightly endothermic, and it thereby requires addition of external heat energy to be sustained. While reaction kinetics demands that the reaction represented by Equation [7.2] be conducted at high temperatures to efficiently dehydrogenate ammonia, reasonably high ammonia decomposition can be achieved at low temperatures (as low as 200°C, which is the temperature at which thermal decomposition of ammonia commences) thermodynamically (Thomas and Parks, 2006). As mentioned in Chapter 6, about 6 kWh of energy is required to release 1 kg of hydrogen from ammonia during its decomposition reaction (Andersson and Grönkvist, 2019). Temperatures over and above 600°C and the use of a catalyst, such as Ru/SiO_2, are required to decompose ammonia completely into hydrogen and nitrogen (Cheddie, 2012). According to Yang et al. (2023), the idea of electrolytically or electro-oxidatively decomposing ammonia was proposed as an alternative dehydrogenation process (in view of their low theoretical thermodynamic energy consumption), which was found not to be promising.

Rather than decomposing ammonia to release hydrogen for power or heat generation, ammonia can be directly combusted in an ammonia-based solid oxide fuel cell to generate power. However, the combustion of pure ammonia in gas turbine is hindered because of its negative combustion characteristics/attributes, such as low laminar burning velocity, high ignition temperature, slow burning speed, and poor flame stability (Yang et al., 2023). Efforts are, however, being made to address these issues and to develop gas turbines that will be able to burn pure ammonia.

7.2.5.3 Storing Hydrogen in Methanol

The third chemical in which hydrogen can be stored and later released when needed is methanol. This is a well-known commercial process that has a history dating back to about 100 years. It basically involves reaction of hydrogen with carbon dioxide (i.e., hydrogenation of carbon dioxide (CO_2)) at a temperature range of about 200–300°C and a pressure ranging from about 5 MPa to about 10 MPa in the presence of Cu-based heterogeneous catalysts, such as Cu/ZrO_2, Cu-ZnO/ZrO_2, etc. (Van der Ham et al., 2012). The hydrogenation of CO_2 to produce methanol that provides storage for hydrogen is represented by Equation [7.3]:

$$CO_2 + 3H_2 \leftrightarrow 2CH_3OH + H_2O \quad \Delta H° = -49.2 \text{ kJ/mol} \qquad [7.3]$$

This phenomenon presents a positive attribute for energy at two fronts – storing and utilizing CO_2 (a greenhouse gas) emitted from various sources (including power generation plants and other industrial sources) to produce methanol (thereby contributing towards addressing the issue of climate change), and by being an energy storage space storing hydrogen (a carbon-free fuel) produced from all sources for later use. The impact on climate change is, however, dependent on the fate of the carbon dioxide produced during the dehydrogenation and/or combustion of methanol.

As depicted by Equation [7.3], the methanol synthesis reaction is an exothermic reaction. As such, the process can be integrated with other processes to produce other valuable industrial products, such as olefins, methyl tert-butyl ether (MTBE), formaldehyde, acetic acid, dimethyl ether (DME), as noted by Yang et al. (2023).

Like in the case with ammonia, methanol can be dehydrogenated to release hydrogen when it is needed. This can be achieved through many ways, among which are the methanol steam reforming and partial oxidation of methanol. Methanol steam reforming is the most favorable route of decomposing methanol to release hydrogen because of its higher hydrogen production per molecule of methanol and relatively lower energy consumption (Behrens and Armbrüster, 2012; Andersson and Grönkvist, 2019). This process is represented by Equation [7.4]:

$$CH_3OH \rightarrow CO + 2H_2 \quad \Delta H^\circ = 90 \text{ kJ/mol} \qquad [7.4]$$

The second pathway from which hydrogen can be released from methanol is by subjecting the methanol to a partial oxidation reaction represented by Equation [7.5]:

$$CH_3OH + 0.5O_2 \rightarrow CO_2 + 2H_2 \quad \Delta H^\circ = -155 \text{ kJ/mol} \qquad [7.5]$$

This is a reaction that takes place when methanol is burned or oxidized under insufficient amount of oxygen or air (sub-stoichiometric air or oxygen) environment.

References

Ahluwalia, R.K., Peng, J.K., Roh, H.S., et al. (2018). Supercritical Cryo-Compressed Hydrogen Storage for Fuel Cell Electric Buses. *Int. J. Hydrogen Energy*, 43, 10215–10231.

Andersson J. and Grönkvist, S. (2019). Large-Scale Storage of Hydrogen. *Int. J. Hydrog Energy*, 44, 11901–11919.

Barron, R.F. (2000). Cryogenic Technology. In *Ullmann's Encyclopedia of Industrial Chemistry*. Ley, C. (Ed.), pp. 481–526. Weinheim: Wiley-VCH Verlag GmbH & Co. KGAA.

Barthelemy, H., Weber, M., and Barbier, F. (2017). Hydrogen Storage: Recent Improvements and Industrial Perspectives. *Int. J. Hydrogen Energy*, 42, 7254–7262.

Behrens, M. and Armbrüster, M. (2012). Methanol Ste am Reforming. In *Catalysis for Alternative Energy Generation*. Guczi, L. and Erdôhelyi, A. (Eds.), pp. 175–235. New York: Springer New York.

Brunner, T. and Kircher, O. (2016). Cryo-compressed Hydrogen Storage. In *Fuel Cells – Data, Facts and Figures*. Stolten, D., Samsun, R.C., Garland, N. (Eds.), pp. 162–174. Weinheim, Germany: Wiley-VCH Verlag GmbH & Co. KGaA.

Bünger, U., Landinger, H., Pschorr-Schoberer, E., Schmidt, P., and Weindorf, W. (2014). Power-To-Gas (PtG) in Transport: Status Quo and Perspectives for]Development. A Report to the Federal Ministry of Transport and Digital Infrastructure (BMVI), Germany.

Cheddie, D. (2012). Ammonia as a Hydrogen Source for Fuel Cells: A Review. In *Hydrogen Energy: Challenges and Perspectives*. Minic, D. (Ed.). London: IntechOpen.

Derking, H. (2019). Liquid Hydrogen Storage: Status and Future Perspectives. In *Cryogenic Heat and Mass Transfer (CHMT)*. Enschede, The Netherlands: Oxford University.

Dutta, S. (2014). A Review on Production, Storage of Hydrogen, and its Utilization as an Energy Resource. *J. Ind. Eng. Chem.*, 20, 1148–1156.

Hua, T.Q., Ahluwalia, R., Peng, J-K., Kromer, M., Lasher, S., McKenney, K., Law, K., and Sinha, J. (2010). Technical Assessment of Compressed Hydrogen Storage Tank Systems for Automotive Applications. A Final Report Prepared for the U.S. Department of Energy by Argonne National Laboratory and IIAX LLC.

HyUnder. (2014). Assessment of the Potential, the Actors and Relevant Business Cases for Large Scale and Long-Term Storage of Renewable Electricity by Underground Storage in Europe (executive summary). http://hyunder.eu/wp-content/uploads/2016/01/D8.1_HyUnder-ExecutiveSummary.pdf.

International Energy Agency. (2019). The Future of Hydrogen – Seizing Today's Opportunities. A Report by the IEA for the G20.

Klerke, A., Christensen, C.H., Nørskov, J.K., et al. (2008). Ammonia Hydrogen Storage: Challenges and Opportunities. *J. Mater. Chem.*, 18, 2304.

Kunze, K. and Kircher, O. (2012). Cryo-Compressed Hydrogen Storage. Presented at Cryogenic Cluster Day, Oxford, UK, 28 September 2012.

Leachman, J. (2015). Business Plan of the Company Protium Innovations LLC – Scalable Hydrogen Liquefier. https://s3.wp.wsu.edu/uploads/sites/44/2015/02/EIC-Business-plan_PMA.pdf.

Leinweber, A. and Müller, K. (2018). Hydrogenation of the Liquid Organic Hydrogen Carrier Compound Monobenzyl Toluene: Reaction Pathway and Kinetic Effects. *Energy Technol.*, 6(3), 513–520.

Matveev, K.I. and Leachman, J. (2021). Numerical Simulations of Cryogenic Hydrogen Cooling in Vortex Tubes with Smooth Transitions. *Energies*, 14, 1429–1442.

Moreno-Blanco, J., Petitpas, G., Espinosa-Loza, F., et al. (2019). The Storage Performance of Automotive Cryo-compressed Hydrogen Vessels. *Int. J. Hydrogen Energy*, 44, 16841–16851.

Parks, G., Boyd, R., Cornish, J., et al. (2014). Hydrogen Station Compression, Storage, and Dispensing Technical Status and Costs: Systems Integration. NREL/BK-6A10-58564. Golden, CO, USA: National Renewable Energy Laboratory.

Perera, M.S.A. (2023). A Review of Underground Hydrogen Storage in Depleted Gas Reservoirs: Insights into Various Rock-Fluid Interaction Mechanisms and Their Impact on the Process Integrity. *Fuel*, 334(Part 1), 126677.

Rao, P.C. and Yoon, M. (2020). Potential Liquid-Organic Hydrogen Carrier (LOHC) Systems: A Review of Recent Progress. *Energies*, 13, 6040.

Sainz-Garcia, A., Abarca, E., Rubi, V., et al. (2017). Assessment of Feasible Strategies for Seasonal Underground Hydrogen Storage in a Saline Aquifer. *Int. J. Hydrogen Energy*, 42, 16657–16666.

Swanger, A.M., Notardonato, W.U., Fesmire, J.E., et al. (2017). Large-Scale Production of Densified Hydrogen to the Triple Point and Below. *IOP Conf. Series: Mat. Sci. Eng.*, 278, 012013.

Tarkowski, R. (2019). Underground Hydrogen Storage: Characteristics and Prospects. *Renew Sustain Energy Rev.*, 105, 86–94.

Thomas, G. and Parks, G. (2006). Potential Roles of Ammonia in a Hydrogen Economy A Study of Issues Related to the Use of Ammonia for On-Board Vehicular Hydrogen Storage. A Study Report for the U.S. Department of Energy.

Valenti, G. (2016). Hydrogen Liquefaction and Liquid Hydrogen Storage. In *Compendium of Hydrogen Energy*. Gupta, R.B., Basile, A., Veziroğlu, T.N. (Eds.), Volume 2, pp. 27–51. Hydrogen Storage, Transportation, and Infrastructure. Swanston, UK: Woodhead Publishing.

Van der Ham, L.G.J, Van den Berg, H., Benneker, A. et al. (2012). Hydrogenation of Carbon Dioxide for Methanol Production. *Chem. Eng. Transact.*, 29, 181–186.

Xu, Z., Yan, Y., Wei, W., et al. (2020). Supply System of Cryo-Compressed Hydrogen for Fuel Cell Stacks on Heavy Duty Trucks. *Int. J. Hydrogen Energy*, 45, 12921–12931.

Yan, J., Wang, W., Miao, L., et al. (2018). Dehydrogenation of Methylcyclohexane Over Pt Sn Supported on Mg Al Mixed Metal Oxides Derived from Layered Double Hydroxides. *Int. J. Hydrogen Energy*, 43, 9343–9352.

Yang, M., Hunger, R., Berrettoni, S., Sprecher, B., and Wang, B. (2023). A Review of Hydrogen Storage and Transport Technologies. *Chem Energy*, 7(1), 190–216.

Zivar, D., Kumar, S., and Foroosesh, J. (2021). Underground Hydrogen Storage: A Comprehensive Review. *Int. J. Hydrogen Energy*, 46, 23436–23462.

8

Hydrogen Utilization

8.1 Introduction

Hydrogen has the potential to be utilized and applied in a diverse and wide range of sectors, including the power generation, industrial manufacturing (chemical, fertilizer, iron and steel, cement), oil refining and processing (including fuel upgrading), synthetic crude/fuel production, transportation, energy storage and carrier (to improve the economics of existing and emerging electric power generation systems), process heat, and residential/commercial building sectors. Its present use is largely found in the transportation, power generation, and industrial manufacturing sectors (mostly oil refining, iron and steel, and chemical industries). Furthermore, hydrogen-enriched syngas can be used to make gasoline and diesel fuel.

This chapter presents the various ways in which hydrogen is used in the various sectors listed before. To effectively utilize hydrogen like all fuels, the energy contained in hydrogen must be converted into another form, depending on what the hydrogen is to be used for, such as electricity, heat, or another form of energy or manufacturing product. This conversion can be achieved through combustion using turbines, process furnace, boiler furnace, or reciprocating engines, or through an electrochemical process using a fuel cell, or through some other combustion devices such as domestic gas burners for cooking or heating. As such, this chapter first discusses some fundamentals of hydrogen combustion and reactions that enable its effective utilization.

8.2 Hydrogen Utilization in Combustion System

Hydrogen combustion is a process where hydrogen is chemically reacted with an oxidizing agent (such as air or oxygen) that gives off heat energy that can be used to generate electricity, propel transportation engines, manufacture various products, and provide comfort in residential and commercial buildings. The only other product of hydrogen combustion is water vapor or steam. This phenomenon enables the ability to generate electricity, manufacture chemical products, and produce heat energy without the emission of carbon dioxide (CO_2). This is achieved by reacting

DOI: 10.1201/9781003348283-8

hydrogen with either air or oxygen at high temperatures. This reaction is represented by Equation [8.1].

$$2H_2 + O_2 \rightarrow 2H_2O + \text{Heat (286 kJ/mol)} \qquad [8.1]$$

As indicated, the reaction (Equation 8.1) is strongly exothermic and it, therefore, gives off about 286 kJ/mol of heat energy. While the reaction depicted in Equation [8.1] is the overall reaction of hydrogen with air/oxygen, there are several intermediate reaction steps before arriving at the overall reaction. Many studies have been conducted, including those conducted by Marinov et al. (1996), Christiansen et al. (2001); Law (2006), Kim et al. (1994), Miller et al. 1982), and Lu et al. (2001), on the mechanisms of hydrogen–air combustion and the various intermediate reaction steps involved. Kim et al. (1994) developed a detailed hydrogen–air combustion mechanism with 19 reactions. The 19 reaction steps are represented by Equation [8.2] through [8.20].

$$H + O_2 \rightarrow O + OH \qquad [8.2]$$

$$O + H_2 \rightarrow H + OH \qquad [8.3]$$

$$OH + H_2 \rightarrow H + H_2O \qquad [8.4]$$

$$O + H_2O \rightarrow OH + OH \qquad [8.5]$$

$$H_2 + M \rightarrow H + H + M \qquad [8.6]$$

$$O + O + M \rightarrow O_2 + M \qquad [8.7]$$

$$O + H_2 \rightarrow OH + H \qquad [8.8]$$

$$O + OH + M \rightarrow H_2O + M \qquad [8.9]$$

$$H + O_2 + M \rightarrow HO_2 + M \qquad [8.10]$$

$$HO_2 + H \rightarrow H_2 + O_2 \qquad [8.11]$$

$$HO_2 + H \rightarrow OH + OH \qquad [8.12]$$

$$HO_2 + O \rightarrow O_2 \qquad [8.13]$$

$$HO_2 + OH \rightarrow H_2O + O_2 \qquad [8.14]$$

$$HO_2 + HO_2 \rightarrow H_2O_2 + O_2 \qquad [8.15]$$

$$H_2O_2 + M \rightarrow OH + OH + M \qquad [8.16]$$

$$H_2O_2 + H \rightarrow H_2O + OH \quad [8.17]$$

$$H_2O_2 + H \rightarrow H_2 + HO_2 \quad [8.18]$$

$$H_2O_2 + O \rightarrow OH + HO_2 \quad [8.19]$$

$$H_2O_2 + OH \rightarrow H_2O + HO_2 \quad [8.20]$$

The species involved in the 19 intermediate reactions include H, O, OH, O_2, H_2O, H_2O_2, and HO_2. Lu et al. (2001) proposed reduced kinetic scheme of reactions from the detailed mechanism developed by Kim et al. (1994). These kinetic reactions are governed by Equation [8.21] through [8.24].

$$O + H_2 \rightarrow H + OH \quad [8.21]$$

$$H + O_2 \rightarrow O + OH \quad [8.22]$$

$$OH + H_2 \rightarrow H + H_2O \quad [8.23]$$

$$H_2 + M \rightarrow H + H + M \quad [8.24]$$

Equations [8.21] to [8.23] are hydrogen–oxygen chain reactions (also known as shuttle reactions), while Equation [8.24] is the H_2–O_2 recombination reaction.

Combustion behavior or performance of hydrogen as a fuel is determined by the various properties of hydrogen. In general and as discussed in Chapter 2, the properties that contribute to its use as a combustible fuel include wide range of flammability, ignition energy, small quenching distance, high autoignition temperature, high flame speed at stoichiometric ratios, very low density, and high diffusivity. While details of these characteristics/properties of hydrogen can be found elsewhere (Chatterjee et al., 2014; Mazloomi and Gomes, 2012; Laurendeau and Glassman, 1971), a brief discussion of some of these characteristics (mainly flammability and flame properties, energy content, and ignition properties) is presented here.

Flammability limits are the entire range of concentrations of a mixture of flammable vapor or gas (hydrogen in this case) in air over which a flame will occur and travel if the mixture is ignited. The lower flammability limit (LFL) is the smallest concentration of the flammable gas in the mixture that can sustain a flame, while the upper flammable limit (UFL) is the highest concentration of the flammable component in the mixture, above which the mixture cannot burn, and the flame will not be sustained.

The autoignition temperature, or self-ignition temperature, of a substance is the lowest temperature at which it spontaneously ignites in a normal atmosphere without an external source of ignition, such as a flame or spark (Laurendeau and Glassman, 1971). This is the temperature needed to provide the activation energy required for sustainable combustion. The autoignition temperature for hydrogen is reported to be about 858 K, which is higher than that of CNG, and a lot higher than that of gasoline (Chatterjee et al., 2014; Mazloomi and Gomes, 2012; Laurendeau and Glassman, 1971). It takes some time between when a substance (fuel) is exposed to the ignition source and when it auto ignites. This time, which is relatively short and varies with fuel, can be estimated for hydrogen using Equation [8.25] (Quintiere, 1997):

$$t_{ig} = \frac{\pi}{4} k\rho c \left\{ \frac{T_{ig} - T_0}{q} \right\}^2 \qquad [8.25]$$

where t_{ig} is the time it takes for a hydrogen–air mixture to reach its autoignition temperature, k is thermal conductivity, ρ is density, c is specific heat capacity, T_{ig} is the flame temperature of the H_2–air mixture, q is the heat, and T_0 is the initial temperature of the hydrogen.

Flame velocity or burning velocity is defined as the velocity at which unburned gases move through the combustion zone in the direction normal to the flame front. This combustion property is related to the ability of the flame to remain stable during the combustion of the fuel. According to Chatterjee et al. (2014), hydrogen has a higher flame speed than those of CNG and gasoline. This means that hydrogen engines can more closely approach the thermodynamically ideal engine cycle, thereby enhancing the break thermal efficiency. At leaner mixtures, however, the flame velocity decreases significantly.

During combustion, energy is released to the combustion products, thereby raising their temperature; adiabatic flame temperature is, therefore, defined as the temperature of the products of combustion after all chemical reactions have reached equilibrium and when no heat is allowed to escape (or enter) the combustor. Each fuel has a unique adiabatic flame temperature for a given amount of air. Adiabatic flame temperature is usually achieved when the combustion takes place under stoichiometric air/fuel ratio. While it can be measured experimentally, it can be calculated using the equilibrium composition of the products of combustion and the respective specific heat capacities. As reported by Chatterjee et al. (2014), the adiabatic flame temperature of hydrogen is lower than that of gasoline but higher than that of compressed natural gas.

The fuel/air ratio to achieve complete combustion of a fuel is referred to as its stoichiometric ratio. That ratio is 34.5/1 for hydrogen–air combustion, as reported in a paper published by Chatterjee et al. (2014). Operating the combustion system at a fuel/air ratio higher than the stoichiometric ratio is known as fuel-rich condition and usually results in an incomplete combustion, while a fuel/air ratio less than the stoichiometric ratio is referred to as fuel lean condition.

Another important combustion property/characteristic of fuel is the lower heating value (LHV). The LHV (also referred to as net calorific value) of a fuel is defined as the quantity of heat generated by fully combusting a specified quantity of the fuel minus the heat of vaporization of the water in the combustion product. According to Chatterjee et al. (2014), the lower heating value of hydrogen is about 120 MJ/kg, which is much higher than those of CNG and gasoline. Therefore, hydrogen will generate more energy per unit mass of hydrogen during its combustion than per unit mass of CNG and gasoline.

As Mazloomi and Gomes (2012) revealed in their paper, hydrogen has higher flammability range, ignition temperature, and flame speed than CNG and gasoline. The high flammability range of (4–75 percent), reported for hydrogen by Mazloomi and Gomes (2012) signifies that hydrogen flame has the potential to be stable over a wide range of hydrogen/fuel ratio and under highly dilute conditions, thereby enabling better operational flexibility, emission reduction, and fuel consumption (Chatterjee et al., 2014), as discussed in Chapter 2. Consequently, the propensity of engine knocking is relatively less when the engine is fired with hydrogen than with gasoline (for example) because of the relatively high ignition temperature and high flame speed of hydrogen, as mentioned in Chapter 2. This is even more so with higher research octane number exhibited by hydrogen compared with gasoline (>120 vs. 91–99), as the data reported in the literature depict (Chatterjee et al., 2014; Mazloomi and Gomes, 2012). Octane numbers are a measure of the ability of a fuel to cause an engine to knock during operation. Knocking occurs when a secondary detonation fuel occurs inside an engine that leads to a temperature increase over and above the autoignition of the fuel, thereby causing the engine to knock. The higher the octane number of a fuel, the lower is its propensity to result into this unwanted combustion phenomenon. Also, the wide flammability range exhibited by hydrogen offers it a wide range of applications as a fuel (Momirlan and Veziroglu, 2005). Another attribute of wide range of flammability exhibited by hydrogen is that hydrogen can run on a lean mixture, is one in which the amount of fuel is less than the theoretical, stoichiometric, or chemically ideal amount needed for combustion with a given amount of air. This is why it is relatively easy to start a hydrogen engine.

Flash point is the temperature at which a fuel generates enough vapor to result in a flame when mixed with air and in the presence of an ignition source (Astbury, 2008; Hord, 1978). As reported by Chatterjee et al. (2014) and Mazloomi and Gomes (2012), the flash point of hydrogen is very much lower than that of gasoline. As a matter of fact, hydrogen has the lowest flash point among a wide range of common fuels, according to Mazloomi and Gomes (2012). Hydrogen-fired combustion devices are, therefore, expected to be much simpler to start and ignite than those fired with other fuels (Verhelst and Wallmer, 2009).

8.3 Hydrogen Utilization in the Electric Power Generation Industrial Sector

Hydrogen's role in the power generation sector is presently insignificant, accounting for only about 0.2 percent of global electricity produced (IEA, 2019). This is, however, expected to increase in the future as gas turbines and combined-cycle gas turbines (CCGT) become amenable to being fired or co-fired with hydrogen. Increased use of hydrogen-containing fuels (such as ammonia and synthesis gas or producer gas) could further increase the opportunity for hydrogen in the power generation industrial sector.

Application of hydrogen in the power generation sector is diverse—including large-scale power generation, distributed power, combined heat and power (CHP), and backup power (U.S. Department of Energy (DOE) Hydrogen Program Plan, 2020). As alluded to in Chapter 4, electric power can also be produced by electrochemically converting hydrogen using fuel cells or by firing gas turbines with hydrogen (DOE Hydrogen Program Plan, 2020).

Because of the modular nature of fuel cell systems, they are appropriate for variable stationary-power requirements ranging from a few watts to the multi-megawatt scale. Global fuel cell-generated power presently covers a very wide area, including primary and backup power for industrial facilities, businesses, homes, telecommunications towers, data centers, and more (DOE Hydrogen Program Plan, 2020). According to IEA (2019), many residential fuel cells are operating in Japan to provide reliable power and hot water in homes. In the United States, many PEM fuel cells have been deployed for backup power primarily for telecommunications towers, as reported by a December 2019 Open Access Government report. More than 220 MW of stationary fuel cell systems were sold globally in 2018, as reported in the DOE Hydrogen Program Plan (2020).

Another option is to utilize hydrogen-rich gases or hydrogen blended with some other gaseous fuel, such as natural gas, in simple- or combined-cycle power generation system. Gas turbines fired with hydrogen-rich gas or hydrogen/natural gas blends offer such opportunity to generate electric power, as well as provide district heating for residential, commercial, and industrial consumers.

Most hydrogen-derived power generation is not from pure hydrogen. Rather, it is commonly generated from hydrogen-rich product gases such as producer gas from refineries, petrochemical plants, and steel mills. There are, however, some small-scale power generation plants utilizing pure hydrogen as fuel. These include a 12 MW hydrogen-fired CCGT in Italy, while a hydrogen-fired gas turbine is providing heat and electricity to a local community in Kobe, Japan (IEA, 2019). According to Goldmeer (2018), reciprocating gas engines can presently run on a gas mixture containing about 70 percent hydrogen with an expectation of utilizing pure hydrogen in the future.

As summarized in the IEA report (2019), there are four main areas of power generation that hydrogen and hydrogen-based products play some role. These include co-firing with ammonia in existing coal-fired power plants, flexible electric power production, backup and off-grid power generation, and large-scale long-term energy storage. While the large-scale long-term energy storage option has been discussed in Chapter 6, the first three areas will be discussed in this chapter. As mentioned, co-firing of hydrogen with ammonia has been demonstrated in a coal-fired power plant in Japan, its deployment is, however, yet to occur (IEA, 2019). On the other hand, there are over 300,000 fuel cells (producing about 1.6 GW of electricity) and some hydrogen-rich gas-fired commercial gas turbines currently being used for flexible power generation (IEA, 2019). Contribution to the backup and off-grid electricity supply is presently limited to village electrification and combined fuel cell-storage systems. While there is potential for an increased role of hydrogen and hydrogen-rich products for power generation in the future, there are, however, some challenges. These include availability of low-cost and low-carbon hydrogen and hydrogen-based products (such as ammonia), high initial capital costs, competition from other flexible power generation options, and high conversion costs, to mention a few (IEA, 2019).

8.3.1 Co-Firing Ammonia in Coal-Fired Power Plants

In co-firing ammonia with coal, the ammonia may be used directly as fuel mixed with coal or it (ammonia) could be cracked or decomposed into its components (hydrogen and nitrogen) and the released hydrogen is co-fired with coal in the combustion system. The gas (either direct ammonia or hydrogen produced from ammonia) may be pre-mixed with the coal or fired separately through a separate feed system, depending on the combustion technology utilized. In the case of a fluidized coal-fired system, the gas may be used as the fluidizing gas in the fluidized-bed combustion reactor.

While co-firing of ammonia in a commercial coal power plant has been successfully demonstrated in Japan by the Japanese Chugoku Electric Power Corporation (Muraki, 2018), it is, however, yet to be deployed IEA, 2019). This may be due, in part, to the fact that the economics of substituting coal with ammonia depend on the availability of low-cost ammonia, as well as the concerns about an increase in NO_x emissions and potential for ammonia slip into the exhaust gas when ammonia is used as fuel (IEA, 2019). According to the IEA report (2019), it is predicted that about 20 percent of coal-fired power plants in the world could be co-firing ammonia by 2030, which would result in annual ammonia demand of 670 million tons, which would consequently require hydrogen demand of about 120 million tons, as well as reducing CO_2 emission from these coal-fired power plants by about 1.2 giga tons, if the ammonia was produced from low-carbon hydrogen source.

8.3.2 Flexible Electric Power Generation

As the name suggests, a flexible electric power generation system is expected to be amenable to varying electric demand and scale. It may also be expected to accommodate a variety of fuels and operating conditions. Gas turbines and CCGT systems offer this opportunity for hydrogen as a fuel in the range of as low as 3–5 percent hydrogen to as high as 30 percent or higher (IEA, 2019). It is expected that turbines that could accommodate pure hydrogen would be available by 2030 (IEA, 2019).

In addition to hydrogen, ammonia and fuel cells also have the potential to provide flexible electric power. For example, micro gas turbines fired with ammonia have been successfully demonstrated with a power capacity of up to 0.3 MW (Shiozawa, 2019). While the use of larger ammonia-fired gas turbines operating above 2 MW capacity has also been successfully demonstrated, they are, however, still plagued with the slow reaction kinetics encountered by ammonia–air combustion, flame instability, and increased NO_x emissions potential (Valera-Medina et al., 2018). These issues with direct combustion of ammonia are being resolved by using hydrogen produced from ammonia decomposition in the gas turbine combustion system rather than the ammonia (IEA, 2019). As outlined in Chapter 6 and depicted in Equation [8.26]), decomposition or cracking of ammonia to hydrogen and nitrogen is an endothermic reaction that requires about 46 kJ/mol of external source of energy, which can be supplied by gas turbines.

$$NH_3\,(g) \rightarrow 0.5N_2\,(g) + 1.5H_2\,(g) \quad \Delta H = +46\ kJ/mol \qquad [8.26]$$

Another potential possibility for flexible electric power generation could be provided by fuel cells.

8.3.2.1 Fuel Cells

Fuel cells operate by converting the chemical energy of fuels such as natural or synthetic gas and hydrogen to electricity and heat energy. In the case of hydrogen fuel cells, water is the only by-product emitted, and no pollutants such as carbon dioxide and oxides of nitrogen (NO_x) are produced because hydrogen does not contain carbon or nitrogen, nor does hydrogen fuel cell operation involve the use of nitrogen-containing reactant (such as molecular nitrogen) that can result into NO_x emission. Fuel cells can be more efficient than internal combustion engines (ICEs) because the electrochemical reactions in a fuel cell generate electricity directly, whereas in ICE, the chemical energy in the fuel must first be converted into mechanical energy and then into electrical energy, as alluded to in the DOE Hydrogen Program Plan (2020). Fuel cell efficiencies of over 60 percent are said to be achieved (Lohse-Busch et al., 2018), while efficiencies of over 80 percent are reported (U.S. DOE Hydrogen and Fuel Cells Program Record, 2011) to be achievable when fuel cells are integrated with CHP systems.

Although fuel cells are basically like batteries, which consist of electrodes separated by an electrolyte or membrane, they, however, do not need to be recharged like batteries do, and can, therefore, operate for as long as fuel and air are available (DOE Hydrogen Program Plan, 2020). Advantages of fuel cells over combustion engines include (DOE Hydrogen Program Plan, 2020):

- lack of moving parts,
- quiet,
- require no oil changes, and need only minimal maintenance, and
- easily scalable, as individual cells can be stacked together to provide a wide range of power.

These fuel cells, which can generate both electricity and heat from hydrogen at high efficiencies, are known to be popular for flexible operation (IEA, 2019). As mentioned in Chapter 4, there are four main fuel cell technologies – polymer electrolyte membrane fuel cells (PEMFCs), phosphoric acid fuel cells (PAFCs), molten carbonate fuel cells (MCFCs), and solid oxide fuel cells (SOFCs) that are amenable for stationary power generation (IEA, 2019). PEMFCs, which operate at relatively low temperatures (<100°C) and are used for power in micro co-generation units, require pure hydrogen to effectively operate, according to IEA (2019). PAFCs, which use phosphoric acid as electrolyte, are used to generate both stationary electric power and heat for space and water. While PEMFCs operate at low temperatures (<100°C), MCFCs and SOFCs operate at relatively higher temperatures (600°C and 800–1000°C, respectively), which allow them to run on different hydrocarbon fuels without the need for an external reformer to produce hydrogen first. MCFCs are used in the megawatt range for power generation (because of their low power density, resulting in a relatively large size). The produced heat can be used for heating or cooling purposes in buildings and industrial applications. SOFCs have similar application areas, often at smaller scale in the kiloWatt range, such as micro co-generation units or for off-grid power supply.

While global installed fuel cell capacity has been increasing rapidly for some time, the use of hydrogen as fuel in this sector has been limited. For example, only about 4 percent of 1.6 GW of stationary electric power generated by fuel cells in 2018 was obtained from hydrogen, according to an IEA report (2019). Efforts in research and pilot-scale development are being made to co-fire hydrogen and coal in gas turbine for power generation in Japan, and to convert an existing CCGT from natural gas to hydrogen in the Netherlands (Northern Netherlands Innovation Board, 2017). Similarly, a few projects, which include a 30 MW electrolyzer plant, an ammonia production plant, a 10 MW hydrogen-fired gas turbine, and a 5 MW hydrogen fuel cell, are being conducted in Australia under the Lincoln Green Hydrogen effort (IEA, 2019) to meet the electricity demand at a local community (Bruce et al., 2018).

Although electric efficiencies of fuel cells can be in the same range as those of CCGTs (around 50–60 percent), current fuel cells can only operate for a relatively shorter lifetime (ranging from about 10,000 hours to about 40,000 hours), with a relatively higher capital cost than gas turbines (IEA, 2019). However, as reported by IEA (2019), the capital cost of hydrogen fuel cells is expected to go down for PEMFCs unit (Bruce et al., 2018).

8.3.3 Backup and Off-Grid Electric Power Generation

These are mainly standalone power systems that rely mainly on diesel-fired generators to provide electricity for a local community or remote area. Such a system, which can also be used to provide backup power and off-grid electricity, can cost as much as about $0.4/kWh due to the cost of transporting diesel to the communities (IEA, 2019). According to the IEA (2019) report, approximately 2500–3000 such systems were deployed in 2018. Another promising market for backup and off-grid electricity is in the telecommunication industry, which is rapidly expanding, especially in the developing and emerging economies. For an example, 1300 of the 650,000 telecommunication towers currently in India power their towers with diesel-fired generators consume about 1.25 billion gallons of diesel per year, producing about 1.25 million tons of CO_2 per year (Lele, 2019). Hydrogen fuel cells are promising candidates to replace these diesel-fired generators.

For such an application, the hydrogen fuel cell system will consist of hydrogen generation, low-pressure storage (including batteries for short-term frequency control), and a fuel cell (Bruce et al., 2018). Bottled hydrogen can be supplied to remote locations where limited alternative transportation/delivery infrastructure is limited or lacking. Compared with its competing systems that rely solely on batteries, hydrogen has the added advantage of being more cost-effective at scale, longer storage periods, and is capable of withstanding harsher operating conditions and environments commonly experienced in remote locations (Bruce et al., 2018).

Backup electricity can also be provided by fuel cells for power outages and they can provide much-needed electricity for off-grid villages, estates, schools, clinics, communities, and maternity hospitals. For example, a small rural village of 34 households was electrified in 2014 in a trial project in South Africa through a mini grid, relying on electricity supply from three 5 kW methanol fuel cells in combination with a 14 m^3 methanol tank and a 73 kWh battery bank (IEA, 2019). Other similar projects utilizing fuel cells to provide backup power in South Africa have been reported to have sprung up, including the Kwa Zulu Natal province project that provides energy to more than 500 households in 2 rural villages in the area, a fuel cell system installed at a clinic in Gauteng province to provide backup power and cater for power outage for a clinic, and a hydrogen-fuel cell system to provide basic electricity supply for a small community in the Eastern Cape province of South Africa, as further stated in the IEA (2019) report.

An unreliable supply of electricity that causes blackouts is a common problem in many other parts of Africa. Many villages and rural areas in Africa do not even have access to electricity. For example, according to estimates by the World Bank, about 55 percent of Nigerian households nationwide had access to electricity in 2020, while only about 25 percent of residents in rural areas have access to electricity during the same period (U.S. DOE Energy Information Agency, 2023).

8.3.4 Integrated Hybrid Energy System

In addition to the various options (co-firing with ammonia in existing coal-fired power plants, flexible electric power production, backup and off-grid power generation) for generating electric power from hydrogen and hydrogen-rich gases discussed previously, hydrogen also has potential to enhance the value of electric power sector through its integration into hybrid energy systems by integrating electricity generation with energy storage, and/or energy conversion technologies to increase capabilities, value, and/or cost-effectiveness of the overall system. Integrating hydrogen technologies in a hybrid energy system (HES) provides unique attributes in both on-grid and off-grid electric power applications (mid-to-long-term/seasonal energy storage, Hunter et al., 2020), grid leveling and stabilization that enhance the fast-acting dynamic response of electrolyzers (Kurtz et al., 2017), and the ability to co-produce electricity as well as other products such as hydrogen, other hydrogen-based fuels, chemicals, or other products for use in diverse markets, as reported in the DOE Hydrogen Program Plan (2020). Described further are some examples of HES.

8.3.4.1 Integration Grid and Renewable Hybrid Systems

This is a system where a renewable energy-based hydrogen production process such as through electrolysis is integrated with a grid electricity production process such as hydrogen fuel cells or hydrogen-fired gas turbine electric power plant. In this integrated hybrid system, the hydrogen produced by the electrolyzer is used as fuel for either a fuel cell electric power system or a gas turbine to generate electricity. The electricity produced via either the fuel cell or the gas turbine can subsequently be used to produce hydrogen whenever it is needed. According to Eichman et al. (2019), techno-economic analysis will be essential to identify optimal designs for integrating electrolyzers with renewables in grid, microgrid, and off-grid applications.

8.3.4.2 Integrated Fossil–Energy Hybrid Systems

This is a hybrid energy system whereby a fossil-based hydrogen production process (such as natural gas steam reforming or coal gasification) is integrated with a carbon capture and storage (CCS) system, a hydrogen-based electric generation system (such as hydrogen fuel cell), and a carbon dioxide utilization system (such as for enhanced oil recovery or for urea production). Urea (NH_2CONH_2), which is typically produced via the Bazarov reaction, in which carbon dioxide and ammonia are first converted to ammonium carbamate (Equation [8.27]), which is subsequently dehydrated to form urea (Equation [8.28]) (Zhang et al., 2021), has a vital role in producing nitrogen-based fertilizer that is essential for food production.

$$CO_2 + 2NH_3 \rightarrow NH_2COONH_4CO_2 + 2NH_3 \rightarrow NH_2COONH_4 \quad [8.27]$$

$$NH_2COONH_4 \rightarrow NH_2CONH_2 + H_2O \quad [8.28]$$

Carbon-dioxide-enhanced oil recovery (CO_2-EOR) is a tertiary oil recovery method that involves the injection of carbon dioxide into an oil reservoir to reduce molecular weight of the oil, lower minimum miscibility pressure and the viscosity of the oil, thereby improving oil mobility through the well bore. According to DOE Hydrogen Program Plan (2020), pilot-scale plants have been deployed that integrate systems for hydrogen production via steam methane reforming of natural gas with vacuum-swing adsorption to co-produce hydrogen for petroleum refining as well as removing concentrated CO_2 in the product stream for use of enhanced oil recovery in nearby oil fields. The DOE Hydrogen Program Plan (2020) further noted that large-scale gasification plants co-fired with coal, biomass, and waste plastics, could also be integrated with thermal storage, hydrogen production and utilization technologies, carbon capture and utilization. The development of poly-generation systems that use high-temperature fuel cell technologies to produce electricity, heat, and hydrogen from natural gas or coal-derived gas, or biomass- or waste-derived gas, is reported to be under way (DOE Hydrogen Program Plan, 2020).

8.3.4.3 Integrated Nuclear Hybrid Systems

Heat from nuclear power plants can be utilized to produce steam required for steam reforming of natural gas to produce hydrogen. In addition, hydrogen production at the nuclear plant may serve as a source of an additional revenue stream. According to U.S. Department of Energy (2018), research is being conducted to develop materials for SOEC electrolysis that are amenable to the operating temperatures of nuclear energy heat sources. These have attracted growing interest in integrating hydrogen production at nuclear power plants. As reported in the U.S. Department of Energy Hydrogen Program Record #19009 (2020), a 1 GW nuclear power plant can produce about 41,000 metric tons of hydrogen per year, assuming the plant produces hydrogen for 26 percent of the time. The concept for the integrated nuclear hybrid system will be like that described earlier for integrated fossil energy hybrid system except that in the case of the nuclear system, nuclear power will be used, as further enumerated in the U.S. Department of Energy Hydrogen Program Record #19009 (2020) report. It is worth noting that commercial nuclear hybrid systems are yet to be deployed, pilot projects are, however, currently on-going (DOE Hydrogen Program Plan, 2020).

8.4 Hydrogen Utilization in Oil Refining Sector

Oil refining sector, which converts crude oil into various petroleum products (such as liquified petroleum gas (LPG), gasoline, jet fuels, diesel fuel, fuel oil, petroleum coke, and feedstock for the petrochemical industry), accounts for about one-third of total worldwide hydrogen consumption (IEA, 2019). Hydrogen is primarily used in the pretreatment of oil fractions prior to further refining, to remove or reduce the impurities (such as sulfur and nitrogen) through hydrotreatment processes such as hydrodesulfurization and hydro-denitrogenation, and during some refining process steps (such as hydro-cracking, hydro-reforming) to maximize the production of oil-refined products. The demand for hydrogen is expected to grow as regulations for sulfur content of refined oil products become stricter, thereby increasing market potential for hydrogen (IEA, 2019).

Hydrotreatment and hydrocracking are the biggest hydrogen-consuming processes in the refinery (IEA, 2019). As mentioned, about 15–30 percent of the hydrogen demand of a refinery is produced within the refinery via processes such as catalytic reforming of naphtha (U.S. Department of Energy, Energy Information Administration, 2021). While hydrotreating is used to remove or reduce sulfur and nitrogen from the petroleum refined products in order for the refinery to produce clean transportation fuels that will meet the ever increasing stringent regulation for sulfur and nitrogen, as well as to prevent poisoning of the catalysts that are used in the catalytic refining processes, hydrocracking process utilizes hydrogen to upgrade heavy residual oils into lower-molecular-weight, higher-value oil products, such as light and middle distillate that are growing in demand (IEA, 2019). Also, producer gas, which is a hydrogen-rich refinery gas product stream, is utilized as fuel to generate process heat for refinery use.

While hydrogen is utilized by refineries, it is also a product/by-product of some refinery operations. These include the catalytic naphtha reforming (a process that produces high-octane gasoline blending components) and steam cracking. However, the hydrogen produced at the refinery is, in most cases, not enough to meet the refinery's hydrogen demand. According to the IEA (2019) report, the rule of thumb is that about one-third of a refinery hydrogen demand is met by the amount of hydrogen produced at the refinery, the balance of which must be obtained outside of the normal refinery operations (such as via merchant purchase or via a dedicated hydrogen production facility). This option is prone to be plagued with CO_2 emission, supply, and market volatility/vulnerability issues, and consequently price instability. At the same time, there has been significant growth in the demand for hydrogen at the refineries necessitated by growing refining activity and increasing requirements for hydrotreating and hydrocracking.

Hydrogen production (for refinery use) outside of normal refining operations accounts for about 230 million tons of CO_2 ($MtCO_2$) emission per year emissions, which is around 20 percent of total refinery emissions (IEA, 2019). The two main ways to address the CO_2 emission issue (hydrogen production by electrolysis and integration of carbon capture and storage technology with fossil-based hydrogen production) are currently not cost-competitive.

The sulfur compounds/elements typically found in petroleum crude oil include sulfur mercaptans, sulfides, disulfides, polysulfides, thiophenes, benzo-thiophenes. Nitrogen compounds (such as pyridine, anthracene, phenanthrenes) are typically found in most crude oils. The main nitrogen compounds known to be present in oil have been classified into three main groups of compounds – non-heterocyclic, non-basic heterocyclic, and heterocyclic (Feng-Fang and Yen, 1981; Ogunsola, 2006). Examples of non-heterocyclic nitrogen compounds include aniline and pentylamine, while non-basic heterocyclic compounds include pyrrole, indole, and carbazole, and those of basic heterocyclic compounds include pyridine, quinoline, and indoline (Ogunsola, 2006).

8.4.1 Fundamentals of Hydrotreatment of Oil

During oil refining, the crude oil is first passed into an atmospheric distillation unit, where it is separated into different fractions (gases, gasoline/naphtha, jet fuel/kerosene, diesel oil, gas oil, and residual bottoms) according to their boiling temperatures. The residual bottom from the atmospheric distillation unit is sent into a vacuum distillation unit to produce more gas oil (known as vacuum gas oil) and vacuum residuum. The naphtha/gasoline and diesel oil fractions are passed into hydrotreaters to remove unwanted impurities, which, according to Ortega (2021) and Fahim et al. (2010), include nitrogen, organic sulfur compounds, aromatics, olefins, oxygen, and metals of the fractions prior to further processing or use.

Sulfur and nitrogen can poison and deactivate catalysts used in some refining processes and can lead to emission of a criteria pollutant (sulfur dioxide and nitrogen oxides, Fahim et al., 2010), if not removed. The discussion of hydrotreating in this book will be limited to the removal of sulfur and nitrogen, referred to as hydrodesulfurization and hydro-denitrogenation, respectively. As mentioned earlier, hydrodesulfurization is a high-pressure and high-temperature process that uses hydrogen gas to reduce the sulfur in oil fractions (especially gasoline and diesel) to hydrogen sulfide, which is then readily separated from the fuel. The process is based on the nature of the key physicochemical process used for sulfur removal and is one of the most developed and commercialized technologies that catalytically converts organosulfur compounds into stable and environmentally friendly compounds (Babich and Moulijn, 2003; Song, 2003).

Hydrotreating basically involves reacting these oil fractions with hydrogen in three major processing steps – heat exchanger system, hydrotreater reactor unit where the actual hydrotreating occurs, and a stripping unit where the clean product stream (such as the desulfurized/denitrogenated product) is separated out. As described by Ortega (2021), during the hydrotreating process, the feed is first pressurized with hydrogen and passed through a heat exchanger system where it is heated to about 290–430°C prior to being fed into the fixed-bed hydrotreating reactor, where hydrogenolysis and mild hydrocracking reactions occur at a pressure range of 150–250 psig in the presence of a catalyst, thereby converting sulfur, nitrogen, oxygen, and other contaminants to hydrogen sulfide, ammonia, water vapor, and other stable by-products, as appropriate. The catalyst used in the reactor unit is dependent on the contaminant and final products of interest. For sulfur removal, cobalt–molybdenum catalysts are preferred, while nickel–molybdenum catalysts are the catalysts of choice for denitrogenation (Ortega, 2021). The reactor operating conditions (temperature and pressure) also depend on the fraction being hydrotreated. The basic hydrotreatment process for naphtha and diesel oil fractions is briefly described later in this section.

Hydrodesulfurization (HDS) can simply be described as a process in which the organic sulfur species are converted to H_2S and the corresponding hydrocarbon, as represented by the following general reaction (Equation [8.29]):

$$R\text{–}SH + H_2 \rightarrow R\text{–}H + H_2S \quad [8.29]$$

where R represents a hydrocarbon, or specifically by Equations [8.30] and [8.31] (Ortega, 2021):

$$CH_3\text{–}CH_2\text{–}CH_2\text{–}CH_2\text{–}CH_2\text{–}SH + H_2 \rightarrow C_5H_{12} + H_2S \quad [8.30]$$

$$CH_3\text{–}CH_2\text{—}S\text{—}CH_2\text{–} + 3H_2 \rightarrow 2C_3H_8 + 2H_2S \quad [8.31]$$

Hydro-denitrogenation (HDN) involves hydrogenolysis of the strong C–N bonds. This process is based on high severity catalytic hydrotreating that requires large amounts of hydrogen, thereby rendering the process very expensive. HDN of a nitrogen-containing compound to form ammonia is represented by Equation [8.32] (Ortega, 2021):

$$C_7H_7N + 5H_2 \rightarrow C_6H_6 + C_3H_8 + NH_3 \qquad [8.32]$$

HDN of nitrogen compounds (such as pyridine or quinoline) at industrial scale is carried out at high temperatures (300–500°C) and high pressures (up to 200 atm) on the surfaces of solid catalysts such as nickel–molybdenum/alumina (Al_2O_3) and cobalt–molybdenum/Al_2O_3 (Furimsky and Massoth, 2001; Prins, 2001; Ho, 1988; and Katzer and Sivasubramanian, 1979). The elementary steps in the industrial HDN process were thought to include sequentially *N*-heterocycle coordination to surface metal sites, aromatic ring hydrogenation, ring opening through C–N bond breaking to yield a linear amine, and denitrogenation to obtain nitrogen-free linear hydrocarbons with the release of ammonia, as shown in Equation [8.33] for HDN reaction mechanism of pyridine (Furimsky and Massoth, 2001; Prins, 2001).

N → N–H → NH₂ → NH₃ [8.33]

The complexity of the solid catalysts, which may contain different types of active sites, however, necessitates the need for further research on the dominant reaction mechanism at the molecular level (Hu et al., 2017).

As noted earlier, the design and operation of a hydrotreater depends on the feedstock and product of interest. The feed oil cuts of interest in this book are naphtha and diesel oil. Described later are hydrotreatment processes for the two refinery product streams.

8.4.2 Naphtha Hydrotreatment Process

Naphtha, a major component of gasoline, is in the C_5–C_{12} range of petroleum refining with a boiling point of the order of 30°C. Light naphtha has an initial boiling point (IBP) of about 30–205°C (Ortega, 2021). As described by Rao (2018), the naphtha is fed into a bank of heaters where it is first mixed with hydrogen and the mixture is heated to 340°C, and the preheated mixture is subsequently fed into the hydrotreatment reactor unit where the hydrotreatment occurs at 315°C and 370 psig pressure in the presence of appropriate catalyst, depending of the targeted impurity (cobalt–molybdenum for sulfur and alumina catalyst bed for nitrogen). After being cooled and condensed, the hydrotreater effluent is passed through a flash separation unit operated at a pressure of 290 psig to separate out the product of interest, which is subsequently sent into a stripping unit operated at 340°C and 305 psig (Ortega, 2021).

8.4.3 Diesel Oil Hydrotreatment Process

Either atmospheric distillation-derived diesel oil or a mixture of fluid catalytic cracker-derived and delayed coking-derived diesel oil can be hydrotreated to reduce 1–2 percent sulfur typically contained in these streams to about 10–15 ppm level required by regulation for diesel fuel (Ortega, 2021). The process steps involved in hydrotreating diesel oil fractions is similar to those for the naphtha except for the addition of the regenerative amine system during hydrotreatment of diesel oil, which recovers excess hydrogen gas and removes hydrogen sulfide using diethanolamine (DEA), thereby making hydrotreatment of diesel oil fractions much more complex than that of naphtha (Ortega, 2021).

8.5 Hydrogen Utilization for Synthetic Crude Oil Production

In addition to conventional crude oil, crude oil can also be produced synthetically from coal, oil shale, and oil sands. Of these three sources by which synthetic crude oil can be produced, only production of coal-derived crude oil involves the use of hydrogen. There are three main technological pathways – pyrolysis, direct liquefaction, and indirect liquefaction – through which crude oil can be produced from coal. The technological pathway that primarily utilizes hydrogen directly is via direct liquefaction technologies, while indirect liquefaction technologies use a mixture of hydrogen and carbon monoxide (known as syngas) produced via other coal conversion processes such as gasification. Hence the discussion in this section is focused mainly on direct coal liquefaction technologies, which utilize hydrogen directly.

The conversion of coal (a solid with atomic carbon/hydrogen (C/H) ratio of approximately 1/1) to oil (a liquid with atomic C/H ratio of roughly 1/2) is, in essence, a process of coal hydrogenation. During direct hydrogenation of coal to produce liquid, the hydrogen may be obtained either directly from gaseous hydrogen or from hydrogen donor solvent. In the former case, the gaseous hydrogen is mixed with pulverized coal slurred with recycled coal-derived liquid in the presence of catalysts (a process known as hydro-liquefaction or catalytic liquefaction developed by Bergius, Berkowitz, 1979), while the latter involving the use of hydrogen donor solvent to liquefy the coal (a process referred to as solvent extraction developed by Pott and Broche; Lee, 1979), as reported by Probstein and Hicks (1982).

The reaction during the direct liquefaction can simply be represented by Equation [8.34] (Probstein and Hicks, 1982):

$$C + 0.8H_2 \rightarrow CH_{1.6} \quad [8.34]$$

The source of the direct hydrogen supply may be through merchant purchase from outside the Syncrude production facility or may be obtained from a hydrogen production process (such as gasification, pyrolysis, partial oxidation, etc.) integrated with the liquefaction process plant. Production of hydrogen from various sources such as electrolysis, fossil fuel, and biomass has been discussed in chapters 3, 4, and 5, respectively. Representing a bituminous coal used for liquefaction by $CH_{0.8}O_{0.1}N_{0.02}S_{0.02}$, and assuming hydrogen is used for liquefaction as well as in complete reduction of all the components to liquid fuel, H_2O, NH_3, and H_2S, respectively, the stochiometric equation for each of the components can be represented by equations [8.35–8.38], respectively (Probstein and Hicks, 1982):

$$CH_{0.8} + 0.4H_2 \rightarrow CH_{1.6} \quad [8.35]$$

$$0.05O_2 + 0.1H_2 \rightarrow 0.1H_2O \quad [8.36]$$

$$0.01N_2 + 0.03H_2 \rightarrow 0.02NH_3 \quad [8.37]$$

$$0.02S + 0.02H_2 \rightarrow 0.02H_2S \quad [8.38]$$

Adding the stochiometric equations [8.35–8.38] reveals that a total of 1.1 unit weight of hydrogen and 15.32 unit weight of coal produces 13.6 unit weight of liquid fuel, as noted by Probstein and Hicks (1982).

As they described, direct coal liquefaction basically involves feeding of pulverized coal and recycled oil or coal-derived solvent slurry mixed with hydrogen into a liquefaction reactor where the coal is liquefied at temperatures ranging from 450 to 475°C, at pressure ranging from 19 to 20 MPa, and at a residence time of about 1 hour in the presence of a catalyst (for hydro-liquefaction) to yield a product stream. The hydrogen required in the case of solvent extraction route is obtained from the donor solvent and from the hydrogen contained in the coal. A simplified flow diagram of generalized direct liquefaction process is shown in Figure 8.1.

The liquefied product stream is sent into a separation unit where it is separated into gas, liquid, and heavy residuum product stream. The liquid product stream, which is typically coal-derived crude oil, is usually heavier and contains higher concentrations

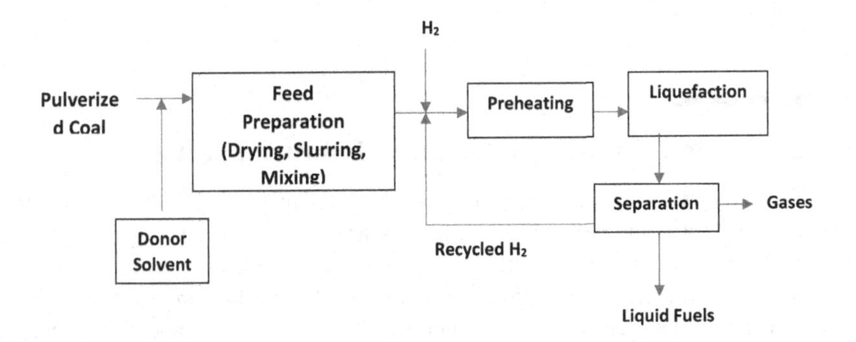

FIGURE 8.1
Simple flow diagram of generalized direct liquefaction process using hydrogen.

of impurities (such as nitrogen, sulfur, ash) than conventional petroleum crude oil. It may, therefore, be necessary to be subjected to some upgrading processes such as hydrogenation, which utilizes hydrogen, prior to being sent to a refinery.

There are two main types of direct liquefaction pathways – solvent extraction and hydro-liquefaction. Examples of solvent extraction process type include solvent refined coal-I (SRC-I) process, solvent refined coal-II (SRC-II) process, and Exxon donor solvent (EDS) process, while the hydro-liquefaction process type includes the H-coal process (which can be operated in either the fuel oil mode or the Syncrude mode) and the Conoco zinc chloride process. More details about these processes and some other coal liquefaction processes can be found elsewhere (Probstein and Hicks, 1982).

In addition to producing coal-derived crude oil through direct liquefaction, an indirect liquefaction technological pathway can be taken. As mentioned earlier, indirect liquefaction involves producing synthesis gas, consisting of mainly carbon monoxide (CO) and hydrogen rather than just hydrogen. In this approach, coal is first converted to syngas via gasification, partial oxidation, or pyrolysis, which is subsequently subjected to a water–gas shift (WGS) reaction to achieve appropriate H_2/CO ratio (typically 2/1 or greater) required to convert the syngas to liquid fuel through a process known as Fischer–Tropsch process. Details of syngas production and subsequent shift reaction have been discussed in Chapter 4. The gasification reaction to produce syngas can simply be represented by Equation [8.39]:

$$C + H_2O \rightarrow CO + H_2 \quad \Delta H = 119 \text{ kJ/mol} \qquad [8.39]$$

This reaction reveals an H_2/CO ratio of 1/1. However, a ratio of at least 2/1 is needed to convert the syngas to liquid. Hence, the reaction shown in Equation [8.39] must be subjected to a WGS reaction to achieve the required ratio of H_2/CO ratio. This can be done by reacting the CO produced in Equation [8.38] with steam to produce more hydrogen represented by Equation [8.40]:

$$CO + H_2O \rightarrow CO_2 + H_2 \quad \Delta H = -40 \text{ kJ/mol} \qquad [8.40]$$

8.6 Hydrogen Utilization in Chemical Industrial Sector

In addition to oil refining and Syncrude production and upgrading, industrial chemical production sector is another sector that utilizes large volumes of hydrogen to produce a complex array of materials, including ammonia, methanol, plastics, fertilizers, solvents, and explosives (IEA, 2019). Of the total 46 million metric tons (Mt) of hydrogen consumed per year by the chemical industrial sector, about 93.5 percent (43 million metric tons) is used to produce ammonia and methanol (IEA, 2019). This is expected to increase to about 57 Mt/year by 2030, as further enumerated by the IEA (2019) report. Hence, the discussion on hydrogen utilization by the chemical sector in this book is restricted primarily to the use of hydrogen to produce ammonia and methanol. A brief mention of hydrogen to produce on ethylene, propylene, benzene, and toluene is attempted.

8.6.1 Hydrogen Use for Ammonia Production

As mentioned in Chapter 7, the Haber–Bosch process is the main industrial ammonia production process, which is carried out by reacting molecular nitrogen (N_2) with hydrogen at 300–500°C and 20–35 MPa pressure in the presence of a metallic catalyst, such as iron (Fe) (Klerke et al., 2008). As discussed in Chapter 7, the process, which is also known as ammonia synthesis, is represented by the following reversible reaction shown in Equation [8.41]:

$$N_2 + 3H_2 \leftrightarrow 2NH_3 \quad \Delta H^\circ = -91.8 \text{ kJ/mol} \qquad [8.41]$$

It is simply a hydrogenation process where nitrogen is hydrogenated to form ammonia. The formed ammonia can then be liquified at −33°C and atmospheric pressure or at 20°C under a pressure of 0.75 MPa (Yang et al., 2023) and stored ready for utilization. For ammonia production at industrial scale, the ammonia synthesis process plant (such as Haber–Bosch process) would be integrated with a hydrogen production system/facility (such as steam methane reforming).

The produced ammonia can be used as a chemical and for power or heat generation by directly combusting the ammonia in either an ammonia-based solid oxide fuel cell or in a gas turbine. However, the combustion of pure ammonia in gas turbine is plagued with some negative combustion characteristics, such as low laminar burning velocity, high ignition temperature, slow burning speed, and poor flame stability (Yang et al., 2023). According to Yang et al. (2023) and Patel (2021), efforts are being made to address these issues and to develop gas turbines that will be able to burn pure ammonia.

8.6.2 Hydrogen Use for Methanol Production

Another chemical that requires a large volume of hydrogen to be produced is methanol. Methanol production from hydrogen is a well-known commercial process that has its history dating back to more than ten decades. As discussed in Chapter 7, it is basically a hydrogenation process whereby CO_2 is reacted with hydrogen at a temperature range of about 200–300°C and a pressure ranging from about 5 to about 10 MPa in the presence of copper-based heterogeneous catalysts, such as Cu/ZrO_2, $Cu\text{-}ZnO/ZrO_2$ (Van der Ham et al., 2012). The hydrogenation of CO_2 to produce methanol is represented by Equation [8.42]:

$$CO_2 + 3H_2 \leftrightarrow 2CH_3OH + H_2O \quad \Delta H^\circ = -49.2 \text{ kJ/mol} \qquad [8.42]$$

The hydrogen-rich syngas used in producing methanol is produced by subjecting natural gas, coal, or biomass to various conversion processes (such as reforming and gasification, as discussed in chapters 4 and 5). During the methanol production process, the hydrogen-rich syngas is compressed to about 5–10 MPa of pressure, and then fed into the methanol synthesis reactor where it is mixed with recycled gas and heated to about 250–270°C, as described by Probstein and Hicks (1982). The effluent of the reactor is then condensed and cooled in a high-pressure separator where the methanol is separated out. The unreacted hydrogen, CO, and CO_2 are tapped off the top of the separator unit.

This concept of producing methanol from carbon dioxide offers two positive attributes – utilizing CO_2 (a greenhouse gas) emitted from various sources (including power generation plants and other industrial sources) to produce methanol (thereby making contributions towards addressing climate change issue) and by producing methanol (a value-added product).

As shown in Equation [8.41], the methanol synthesis reaction is an exothermic reaction, thereby making the process amenable to integration with other processes to produce other valuable industrial chemicals, such as olefins, methyl tert-butyl ether (MTBE), formaldehyde, acetic acid, dimethyl ether (DME), and various solvents, as noted by Yang et al. (2023). Methanol is also used in the production of gasoline through the methanol-to-gasoline (MTG) process.

8.7 Hydrogen Utilization in Iron and Steel Manufacturing Sector

In addition to chemical manufacturing, hydrogen is also used in the manufacturing of other industrial materials such as iron and steel and cement. The discussion in this section is primarily focused on the utilization of hydrogen for steelmaking, which is currently attracting increasing interest. Steel is the most used metal product in the world (DOE Hydrogen Program Plan, 2020). Blast furnace, which utilizes metallurgical coke (usually produced from coking coal) is the conventional method to reduce iron ore to produce steel. In some instances, a mixture of natural gas and coal-derived coke is used as fuel for blast furnace.

Briefly, the basic blast furnace process can be simply represented by the following process steps:

- Combustion of carbon in the fuel to produce heat and carbon dioxide. The heat produced, which is due to the exothermicity of the reaction, is used to support the heat needed for the process. The reaction is represented by Equation [8.43]:

$$C + O_2 \rightarrow CO_2 \quad \Delta H = -406 \text{ kJ/mol} \qquad [8.43]$$

- Decomposition of CO_2 to produce carbon monoxide (CO). In this decomposition zone, the CO_2 produced in the combustion zone is reacted with carbon to produce CO, as represented by Equation [8.44]:

$$CO_2 + C \rightarrow 2CO \quad [8.44]$$

- Reduction of the iron oxides to iron. In this third zone, the CO produced in the decomposition zone is reacted with the iron oxide (e.g., Fe_2O_3) to produce iron (Fe), a reaction which is represented by a simple reaction known as carbothermal reaction, and it is depicted Equation [8.45]:

$$Fe_2O_3 + 3CO \rightarrow 2Fe + 3CO_2 \quad [8.45]$$

The successive processes of sinter belt, blast furnace, and converter used in conventional steel making operate at extremely high temperatures, hence are energy-intensive and produce significant amounts of CO_2 (Li et al., 2021).

Hydrogen can be used as fuel to address the two challenges the conventional coal/coke-fired blast furnace is facing. Hydrogen can, therefore, be used as a fuel to replace the CO_2-ladden coke either partially or fully. In that case the iron oxides reduction reaction shown in Equation [8.45] will now be replaced with Equation [8.46] when iron oxide is reacted with hydrogen:

$$Fe_2O_3 + 3H_2 \rightarrow 2Fe + 3H_2O \quad [8.46]$$

By utilizing hydrogen as the primary reducing agent, the greenhouse gas emissions of about 3000 Mt/year attributed to steel making, which accounts for between 7 and 9 percent of global emissions, can be reduced, according to World Steel Association (2020) and Li et al. (2021). In addition, the direct reduction of crude iron ore by hydrogen does not require the sintering stage, the main energy consumer and CO_2 emission source of the steelmaking process (Li et al., 2021). The use of hydrogen is being demonstrated in many steelmaking plants, including the ones in Austria and Sweden, as noted in the DOE Hydrogen Program Plan (2020). More details about the fundamentals of steelmaking can be found elsewhere (Li et al., 2021; World Steel Association, 2020).

Other manufacturing and chemical industrial sectors utilizing hydrogen include the cement production industry, which is an energy-intensive process that accounts for about 8 percent of world's CO_2 emissions (Lehne and Preston, 2018), and glass manufacturing, food processing industry.

8.8 Hydrogen Utilization in the Transportation Sector

In addition to the industrial applications of hydrogen discussed in sections 8.1–8.7, there is also great potential to utilize hydrogen in all modes of transportation sector (road (e.g., cars, vans, trucks, and buses), rail, air/aviation, and maritime) depending on itis competitiveness with the other transportation fuel options. This section discusses the various ways hydrogen can be utilized as fuel in the transport sector, which relies mostly on oil-derived fuels (gasoline, diesel, jet, and marine) and, except for rail engines, which are electrically powered. More than 33 percent of CO_2 emission in the United States comes from the transportation sector (U.S. DOE EIA, 2019), reduction of which can be addressed by replacing oil-derived transportation fuels with hydrogen-derived fuels, especially in hard-to-decarbonize applications, such as heavy-duty trucks that are used for long-distance hauling, as well as other medium- and heavy-duty vehicles that require longer driving ranges or demand faster refueling times than may be available with just battery electric vehicles (DOE Hydrogen Program Plan, 2020). The medium- and heavy-duty vehicles account for about one-quarter of fuel used every year (Oak Ridge National Laboratory, 2017). The forklift and material-handling section of the transportation sector is an example of an early market success story for hydrogen fuel cells in transportation, which the DOE's early investment in partnership with the private sector helped to achieve (DOE Hydrogen Program Plan, 2020).

Hydrogen utilization for road transportation is currently mostly via fuel cells. According to the IEA (2019) report, there were 4000 fuel cell electric vehicles (FCEVs) sold in 2018. FCEVs tend to have the potential to abate local air pollution since they do not have exhaust emissions. The driving range and pattern of refueling for FCEVs is like ICE vehicles (IEA, 2019). Also, hydrogen is not generally plagued with resource constraints or competition for land use like biofuels do. However, the development of FCEVs has been slow because of associated technical and economic challenges.

In addition to its use in fuel cells, hydrogen can also be used to produce alternative liquid fuels (such as methanol, as described in Section 8.6) to meet the demand for various transportation applications, thereby opening more doors of opportunity for its use in long-distance commercial aircrafts (hard-to-decarbonize system), and ICEs (DOE Hydrogen Program Plan, 2020). According to the U.S. Department of Energy, Office of Energy Efficiency and Renewable Energy (2020), hydrogen is being used in more than 8800 passenger and commercial vehicles, with a growing infrastructure of approximately 45 hydrogen fueling stations in the United States (DOE Hydrogen Program Plan, 2020). It is claimed that there were more than 25,000 fuel cell vehicles and over 470 fueling stations worldwide as of 2019 (IEA, 2020).

The new regulations on sulfur contained in fuel oil used by ocean-going ships reducing its sulfur content from 3.5 to 0.5 percent (The Maritime Executive, 2018), which is further limited to 0.1 percent for ships operating in emission control areas (Merk, 2014), may provide opportunity for hydrogen to replace fuel oils used by ocean-going vehicles (DOE Hydrogen Program Plan, 2020).

Other areas of maritime applications for hydrogen include vessels at ports for drayage trucks, shore power generation equipment, and auxiliary cargo equipment (IEA, 2019). Projects using fuel cells, often in combination with batteries, are planned in California (GGZEM, 2018), and for some Europe-wide operations according to the IEA (2019) report.

The utilization of hydrogen in the rail and aviation industries is currently limited with only two hydrogen trains being in existence, while its use in aviation is only at feasibility stage (IEA, 2019). The most innovative of such technologies are battery electric trains and hydrogen fuel cell trains. To that effect, Germany is expanding her fleet of hydrogen trains. For example, two hydrogen trains that can travel almost 800 km a day on a single refueling are already in use in Lower Saxony, Germany, according to IEA (2019). Other countries such as Austria, the United Kingdom, France, and Japan have since followed the suite (Wiseman, 2019; Kyodo, 2018), as enumerated in the IEA (2019) report. Hydrogen trains could compete with other passenger services options (IEA, 2019). The advantage of utilizing hydrogen in rail could even be greater if it is combined with its use in forklifts, trucks, and other railyard and logistics hub equipment (IEA, 2019).

Aviation, which is responsible for approximately 3 percent of the world's energy-related CO_2 emissions in 2017, presents another area of opportunity for hydrogen utilization (IEA, 2019). Also, changes in aircraft design as well as new refueling and storage infrastructure at airports would be required for the use of pure hydrogen as an aviation fuel because of some characteristic properties of hydrogen (low energy density and the need for cryogenic storage), as outlined in IEA (2019) report. Apart from on-board use of hydrogen in aviation, hydrogen is currently being used for electric power generation (Baroutaji et al., 2019).

8.9 Hydrogen Utilization in the Building Sector

Energy utilization in the building sector is not trivial, accounting for about 30 percent of global energy use mostly for cooking, space heating, and hot water production (IEA, 2019). These utilization areas are presently catered for mostly by fossil fuels and wood fuels. Though challenging and relatively complex and requiring expensive major new infrastructure, hydrogen has the potential to contribute to reducing the reliance on these sources for cooking, heating, and producing hot water in buildings. For these building purposes, hydrogen will serve as a fuel that will be combusted in the building space heating furnace, hot water boiler, and cooking gas burners. The fundamentals of hydrogen, which applies to building devices have been described in Section 8.2. This can be achieved by (1) use of blends of hydrogen with other fuels such as natural gas in existing infrastructure, (2) use of methane produced from hydrogen or hydrogen-containing gas such as syngas, and (3) use of 100 percent pure hydrogen.

According to IEA (2019) report, there are several efforts to demonstrate the feasibility of using hydrogen in residential and commercial buildings such as the 37 demonstration projects aimed at evaluating the feasibility of using hydrogen blending in the gas grid; the H21 Northern England project in the United Kingdom planned to supply about 180 kt/year by 2025 and 2 Mt/year of pure hydrogen by 2035 via pipeline to buildings (Northern Gas Networks, 2018); the micro co-generation and fuel cell hydrogen demonstration projects in Europe and Asia, notably the ENE-FARM project in Japan; and the instillation of more than 1000 small stationary fuel cell systems for residential and commercial buildings in 11 countries in Europe (Ravn Nielsen and Prag, 2017). However, it is worth noting that hydrogen will not be appropriate for all building applications because of the many factors (such as existing natural gas infrastructure, heat densities, safety, etc.) that may influence eventual hydrogen demand in buildings (IEA, 2019).

References

Astbury, G.R. (2008). A Review of the Properties and Hazards of Some Alternative Fuels. *Process Saf. Environ. Prot.*, 86, 397–414.

Babich, I.V. and Moulijn, J.A. (2003). Science and Technology of Novel Processes for Deep Desulfurization of Oil Refinery Streams: A Review. *Fuel*, 82(6), 607–631.

Baroutaji, A., Wilberforce, T., Ramadan, M., and Olabi, A.G. (2019). Comprehensive Investigation on Hydrogen and Fuel Cell Technology in the Aviation and Aerospace Sectors. *Renewable Sustain. Energy Rev.*, 106, 31–40.

Berkowitz, N. (1979). *An Introduction to Coal Technology*. New York: Academic Press.

Bruce, S., Deverell, J., and Hartley, P. (2018). *National Hydrogen Roadmap – Pathways to an Economically Sustainable Hydrogen Industry in Australia*. Australia: CSIRO.

Chatterjee, A., Dutta, S., and Mandal, B.K. (2014). Combustion Performance and Emission Characteristics of Hydrogen as an Internal Combustion Engine Fuel. *J. Aeronaut. Automotive Eng.*, 1(1), 1–6.

Christiansen, E.W., Law, C.K., and Sung, C.J. (2001). Steady and Pulsating Propagation and Extinction of Rich Hydrogen-Air Flames at Elevated Pressures. *Combustion Flame*, 124, 35–49.

Doosan Company. (2018). PureCell Model 400. www.doosanfuelcellamerica.com/download/pdf/catalog/pafc-400kw_us_en.pdf.

Eichman, J., Fernandez, O.G., Koleva, M., and McLaughlin, B. (2019). PG&E H2@Scale CRADA: Optimizing an Integrated Solar-Electrolysis System. National Renewable Energy Laboratory. www.hydrogen.energy.gov/pdfs/htac_nov19_06_eichman.pd

Fahim, M., AL-Sahhaf, T.A., and Elkilani, A.S. (2010). *Fundamentals of Petroleum Refining*. Amsterdam, The Netherlands: Elsevier Science.

Feng-Fang, S. and Yen, T.F. (1981). Concentration and Selective Identification of Nitrogen- and Oxygen-Containing Compounds in Shale Oil. *Anal. Chem.*, 53, 2081–2084.

Furimsky, E. and Massoth, F.E. (2001). Hydrodenitrogenation of Petroleum. *Catal. Rev. Sci. Eng.*, 47, 297–489.

GGZEM (Golden Gate Zero Emission Marine). (2018). Current Projects: The Water-Go-Round. https://watergoround.com/

Goldmeer, J. (2018). Fuel Flexible Gas Turbines as Enablers For A Low Or Reduced Carbon Energy Ecosystem. GE Power, GEA33861, www.ge.com/content/dam/gepower/global/en_US/documents/fuelflexibility/GEA33861%20- %20Fuel%20Flexible%20Gas%20Turbines%20as%20Enablers%20for%20a%20Low%20Carbon%2 0Energy%20Ecosystem.pdf.

Ho, T C. (1988). Hydrodenitrogenation Catalysis. *Catal. Rev. Sci. Eng.*, 30, 117–160.

Hord, J. (1978). Is Hydrogen a Safe Fuel? *Int. J. Hydrogen Energy*, 3, 57–76.

Hu, S., Luo, G., Shima, T., Luo, Y., and Hou, Z. (2017). Hydrodenitrogenation of Pyridines and Quinolines at a Multinuclear Titanium Hydride Framework. *Nature Comm.*, 8, 1866.

Hunter, C., Reznicek, E., Penev, M., Eichman, J., and Baldwin, S. (2020). Energy Storage Analysis. National Renewable Energy Laboratory. www.hydrogen.energy.gov/pdfs/review20/sa173_hunter_2020_o.pdf.

IEA. (2019). The Future of Hydrogen – Seizing Today's Opportunities. Report Prepared by IEA for the G20, Japan.

IEA. (2020). Hydrogen. www.iea.org/reports/hydrogen.

Katzer, J.R. and Sivasubramanian, R. (1979). Process and Catalyst Need for Hydrodenitrogenation. *Catal. Rev. Sci. Eng.*, 20, 155–208.

Kim, T.J., Yetter R.A., and Dryer F.L. (1994). New Results on Moist CO Oxidation: High Pressure, High Temperature Experiments and Comprehensive Kinetic Modeling. In Proceedings of the Twenty-Fifth Symposium (International) on Combustion, The Combustion Institute, pp. 759–766.

Klerke, A., Christensen, C.H., Nørskov, J.K., et al. (2008). Ammonia for Hydrogen Storage: Challenges and Opportunities. *J. Mater. Chem.*, 18, 2304.

Kurtz, J., Harrison, K., Hovsapian, R., and Mohanpurkar, M. (2017). Dynamic Modeling and Validation of Electrolyzers in Real Time Grid Simulation – TV031. www.hydrogen.energy.gov/pdfs/review17/tv031_hovsapian_2017_o.pdf.

Kyodo. (2018). Toyota Teams Up with JR East to Develop Hydrogen-Powered Trains. https://this.kiji.is/418056389112202337.

Laurendeau, N.M. and Glassman, I. (1971). Ignition Temperatures of Metals in Oxygen Atmospheres. *Combust. Sci. Technol.*, 3(2), 77–82.

Law, C.K. (2006). Propagation, Structure, and Limit Phenomena of Laminar Flames at Elevated Pressures. *Combust. Sci. Technol.*, 178, 335–360.

Lee, E.S. (1979). Coal Liquefaction in Coal Conversion Technology. C.Y. Wen and E.S. Lee (Eds.), pp. 428–545. Reading, MA: Addison-Wesley.

Lehne, J. and Preston, F. (2018). Making Concrete Change. Innovation in Low-carbon Cement and Concrete. Chatham House Report. www.chathamhouse.org/sites/default/files/publications/2018-06-13-making-concrete-change-cement-lehne-preston-final.pd.

Lele. (2019). Hydrogen and Fuel Cells at Reliance Industries Limited. The Fuel Cell Industry Review.

Li, S., Zhang, H., Nie, J., Dewil, R., Baeyens, J., and Deng, Y. (2021). The Direct Reduction of Iron Ore with Hydrogen. *J. Sustainability*, 313(16), 8866.

Lohse-Busch, H., Duoba, M., Stutenberg, K., Iliev, S., Kern, M., Richards, B., Christenson, M., and Loiselle-Lapointe, A. (2018). Technology Assessment of a Fuel Cell Vehicle: 2017 Toyota Mirai. Argonne National Laboratory. ANL/ESD-18/12 https://publications.anl.gov/anlpubs/2018/06/144774.pdf.

Lu, T., Ju, Y., and Law, C.K. (2001). Complex CSP for Simplifying Kinetics. *Combustion Flame*, 126, 445–455.

Marinov, N.M., Westbrook, C.K., and Pitz, W.J. (1996). Detailed and Global Chemical Kinetics model for hydrogen. In S.H. Chan, (Ed.), *Transport Phenomena in Combustion*, Vol. 1, pp. 118–129. Washington, DC, USA: Taylor & Francis.

Mazloomi, K. and Gomes, C. (2012). Hydrogen as an Energy Carrier: Prospects and Challenges. *Renewable Sustainable Energy Rev.*, 16, 3024–3033.

Merk, O. (2014). Shipping Emissions in Ports, International Transport Forum, p. 15. www.itf-oecd.org/sites/default/files/docs/dp201420.pdf.

Miller, H.P., Mitchell, R., Smooke, M., and Kee, R. (1982). Towards a Comprehensive Chemical Kinetic Mechanism for the Oxidation of Acetylene: Comparison of Model Predictions with Results from Ame and Shock Tube Experiments. Proceedings of the Nineteenth Symposium (International) on Combustion, p. 181196. Pittsburgh, PA, USA: The Combustion Institute.

Momirlan, M. and Veziroglu, T.N. (2005). The Properties of Hydrogen as Fuel Tomorrow in Sustainable Energy System for a Cleaner Plant. *Int. J. Hydrogen Energy*, 30, 795–802.

Muraki, S. (2018). R&D on Hydrogen Energy Carriers Toward Low-Carbon Society. Presentation on 12 October 2018 at IEA.

Northern Gas Networks. (2018). H21 North of England, https://northerngasnetworks.co.uk/h21-noe/H21-NoE-23Nov18-v1.0.pdf.

Northern Netherlands Innovation Board. (2017). Green Hydrogen Economy in the Northern Netherlands. http://verslag.noordelijkeinnovationboard.nl/uploads/bestanden/dbf7757e-cabc-5dd6-9e97-16165b653dad/3008272975/NIB-Hydrogen-Full_report.pdf.

Oak Ridge National Laboratory. (2017). Transportation Energy Data Book 36. https://info.ornl.gov/sites/publications/Files/Pub104063.pdf.

Ogunsola, O.M. (2006). Denitrogenation of Bitumen/Syncrude by Modified Supercritical Water. A Project Report Submitted to the U.S. Department of Energy Under DOE Grant No. DE-FG02-05ER084217.

Ortega, E. (2021). An Overview of Hydrotreating. *Chem. Eng. Progr.*, October 29–33.

Patel, S. (2021). Mitsubishi Power Developing 100 Percent Ammonia-Capable Gas Turbine. POWER Magazine.

Prins, R. (2001). Catalytic Hydrodenitrogenation. *Adv. Catal.*, 46, 399–464.

Probstein, R.F. and Hicks, R.E. (1982). *Synthetic Fuels*. Chemical Engineering Book Series. New York, NY: McGraw-Hill, Inc.

Quintiere, J.G. (1997). Principles of Fire Behavior. Independence, KY: Delmar Publishers. ISBN 0-8273-7732-0.

Rao, B. (2018). *Modern Petroleum Refining Processes* (6th ed.). New Delhi, India: CBS Publishers and Distributors Private Ltd.

Ravn Nielsen, E. and Prag, C.B. (2017), Learning Points from Demonstration Of 1000 Fuel Cell Based Microchp Units – Summary of Analyses from The Ene Field Project.

Shiozawa, B. (2019). SIP Energy Carriers – Updates and Establishment of Green Ammonia Consortium. Presentation, 26 February 2019. http://injapan.no/wp-content/uploads/2019/02/3-SIP-EnergyCarriers.pdf.

Song, C. (2003). An Overview of New Approaches to Deep Desulfurization for Ultra-clean Gasoline, Diesel Fuel and Jet Fuel. *Catalysis Today*, 86(1–4), 211–263.

The Maritime Executive. (2018). MO Answers Questions on the 2020 SOx Regulation. www.maritime-executive.com/article/imo-answers-questionson-the-2020-sox-regulation.

U.S. Department of Energy. (2018). Energy Department Announces up to $3.5m for Nuclear-Compatible Hydrogen Production. www.energy.gov/eere/articles/energy-departmentannounces-35m-nuclear-compatible-hydrogen-production.

U.S. Department of Energy, Energy Information Administration. (2021). EIA-820 Annual Refinery Report, EIA, Washington, DC, USA.

U.S. Department of Energy, Energy Information Administration. (2023). Country Analysis Brief: Nigeria. International – U.S. Energy Information Administration (EIA).

U.S. Department of Energy, Hydrogen and Fuel Cells Program Record 11014. (2011). Medium-scale CHP Fuel Cell Systems Targets. www.hydrogen.energy.gov/pdfs/11014_medium_scale_chp_target.pdf.

U.S. Department of Energy, Office of Energy Efficiency and Renewable Energy. Alternative Fuels Data Center. (2020). Hydrogen Fueling Station Locations. https://afdc.energy.gov/fuels/hydrogen_locations.html#/find/nearest?fuel=HY

U.S. Department of Energy Hydrogen Program Plan. (2020). DOE/EE – 2128.

U.S. Department of Energy Hydrogen Program Record #19009. (2020). Hydrogen Production Cost from PEM Electrolysis – 2019. www.hydrogen.energy.gov/pdfs/19009_h2_production_cost_pem_electrolysis_2019.pdf

U.S. Department of Energy Information Administration. (19 March 2019). Today in Energy. www.eia.gov/todayinenergy/detail.php?id=38773.

Valera-Medina, A., and Xiao, H. (2018). Ammonia for Power. *Prog. Energy Combust. Sci.*, 69, 63–102.

Van der Ham, L.G.J, Van den Berg, H., Benneker, A., et al. (2012). Hydrogenation of Carbon Dioxide for Methanol Production. *Chem. Eng. Transact.*, (29), 181–186.

Verhelst, S. and Wallmer, T. (2009). Hydrogen-Fueled Internal Combustion Engine. *Prog. Energy Combust. Sci.*, 35, 490–527.

Wiseman, E. (2019). Hydrogen Fuel Cell Trains to Run on British Railways from 2022. *The Telegraph.* www.telegraph.co.uk/cars/news/hydrogen-fuel-cell-trains-run-british-railways-2022/.

World Steel Association. (2020). Steel's Contribution to a Low Carbin Future and Climate Resilient Societies. www.worldsteel.org/en/dam/jcr:7ec64bc1-c51c-439b-84b8-94496686b8c6/Position_paper_climate_2020_vfinal.pdf.

Yang, M., Hunger, R., Berrettoni, S., Sprecher, B., and Wang, B. (2023). A Review of Hydrogen Storage and Transport Technologies. *Chem Energy*, 7(1), 190–216.

Zhang, H., Wang, L., Van herle, J., Maréchal, F., and Desideri, U. (2021). Techno-Economic Comparison of 100% Renewable Urea Production Processes. *Appl. Energy*, 284, 116401.

Index

Printed in the United States
by Baker & Taylor Publisher Services